THE OCEAN'S BODY

MARINE ECOLOGY THROUGH THE LENS OF HUMAN ANATOMY

Albert Calbet

Institut de Ciències del Mar, CSIC, Barcelona. Spain

Albert Calbet
Institut de Ciències del Mar, CSIC.
Copyright © 2024 Albert Calbet
All rights reserved.
ISBN: 9798343294262

CONTENTS

About the author

Introduction

Chapter 1. The Oceanic Origin of Life and the Living Memory of Seawater

Chapter 2. The Skin of the Sea: The Ocean's Surface as a Living Boundary

Chapter 3. The Circulatory System: Ocean Currents as Blood Vessels

Chapter 4. The Digestive System: Zooplankton as the Gut of the Ocean

Chapter 5. The Immune System: Defenses and Toxins

Chapter 6. The Brain and Nervous System of the Ocean

Chapter 7. Chemical Signaling in the Ocean: Prey Detection and the Human Senses

Chapter 8. The Skeleton of the Sea: Siliceous and Calcareous Organisms as Bone Builders

Chapter 9. The Lungs of the Ocean: Plankton and Oxygen Production

Chapter 10. The Reproductive System: Marine Life Cycles and Reproduction

Chapter 11. The Excretory System: Plankton and Nutrient Recycling

Chapter 12. The Hormonal Cycles of the Ocean: Seasonality and Biological Rhythms

Chapter 13. Diagnosing Ocean Illness: A Parallel to Human Diseases

Chapter 14. Oceanic Healing: Treating the Sea's Ailments as Doctors of the Planet

Chapter 15. The Ocean as a Living Being

About the Author

My journey into the world of living creatures began at a young age, sparked by a simple microscope I received as a child. Early observations of microscopic life inspired a lifelong curiosity about the natural world, eventually leading me to study Biology at the University of Barcelona. After completing my degree in 1992, I pursued a PhD at the Marine Sciences Institute (CSIC, Barcelona), focusing on marine zooplankton, a field that would shape much of my career.

Upon completing my PhD, in 1997, I continued my research as a postdoctoral fellow at the University of Hawaii (USA), where I developed a deep interest in marine microzooplankton, a subject that remains central to my work today. My time in Hawaii was a defining period, expanding both my academic horizons and my understanding of marine ecosystems.

Returning to Barcelona, I spent several years on temporary contracts, continuing my research in marine plankton. Eventually, I secured a permanent research position at the Marine Sciences Institute in 2006, where I have worked for over three decades. My research has focused primarily on zooplankton ecology, contributing to a broader understanding of marine ecosystems and their crucial role in the global environment.

Throughout my career, I have remained passionate about making scientific knowledge accessible to a wider audience. Sharing my expertise with the overall public is a personal priority, which has led me to write this book. I hope you enjoy it!

INTRODUCTION

Imagine the ocean as a vast, living being—its currents flowing like blood through arteries, its coral reefs functioning like sensory hubs, its phytoplankton fueling life just as mitochondria power our cells. The ocean is more than just a collection of ecosystems; it is a cohesive, interconnected organism that responds, adapts, and functions as a whole, much like the human body.

At first glance, comparing the ocean to the human body may seem like a stretch of the imagination. After all, we are made of flesh, bones, and blood, while the ocean is an expanse of saltwater teeming with marine life. But when you dive deeper, you begin to see the parallels. Both the human body and the ocean rely on complex systems working in harmony to sustain life. Just as our bodies depend on the coordinated efforts of organs, cells, and tissues to function, the ocean depends on the interplay of ecosystems, organisms, and physical processes to maintain its balance and support life on Earth.

The idea behind *The Ocean's Body* is to explore these parallels and offer a new way of understanding marine ecology by drawing comparisons to human anatomy. Certainly, I could have used the anatomy of any other animal, such as a pangolin, a fish, or a sea urchin. However, we are all more familiar with our own anatomy, and when it comes to raising awareness about the dangers the ocean faces, we tend to feel closer and more empathetic when we relate these issues to diseases that could affect us personally. Who cares if a pangolin has a heart attack? But when it is ourselves who might suffer this condition, the analogy with the ocean becomes much more concerning. Take, for instance, the ocean's circulation system. Much like our blood vessels transport oxygen, nutrients, and waste products throughout the body, ocean currents circulate heat, nutrients, and life across the globe. The Gulf Stream and the Antarctic Circumpolar Current act as the ocean's arteries, delivering essential resources to ecosystems that would otherwise wither and die. Disrupt these currents—just as you might block an artery—and the consequences are catastrophic. Ecosystem's collapse, climate patterns shift, and life in the ocean begins to falter.

The ocean, like the human body, is more than the sum of its parts. It operates as a whole, with each element—plankton, currents, coral reefs, fish, marine mammals—playing an essential role in the broader system. Yet, these elements are not isolated; they interact, adapt, and respond to changes, often in ways that are invisible to us but critical to the ocean's overall health. When one part of the ocean is disrupted, the ripple effects are felt throughout, just as the malfunction of a single organ or system in the body can affect our entire health.

This interconnectedness is not just a scientific curiosity; it is a vital truth about how life on Earth operates. The ocean's metabolic processes, driven by the energy captured by phytoplankton, are the foundation of all marine food webs. These tiny organisms, invisible to the naked eye, act as the ocean's "powerhouses," much like mitochondria fuel our cells. They convert sunlight into energy, sustaining everything from the smallest zooplankton to the largest whales. When their populations decline due to warming waters or pollution, it sets off a chain reaction that affects the entire marine food web. The ocean's "body" begins to fail.

The ocean's surface, akin to the skin of a living organism, acts as a dynamic boundary that regulates Earth's climate, exchanges gases, and sustains marine life. Phytoplankton play a vital role also in carbon sequestration, and regulating sunlight absorption, much like human skin's protective and regulatory functions.

In the human body, we often see how the nervous system coordinates responses to external stimuli, allowing us to sense, react, and adapt to our environment. The ocean, too, has its sensory networks, from the coordinated movements of fish schools that respond in unison to predators, to the long-distance communication of whales that echo across vast stretches of water. These behaviors and interactions are part of the ocean's broader "nervous system," a decentralized web of reactions that help maintain balance and resilience in a dynamic and often hostile environment.

The comparison between the ocean and the human body is not just about finding poetic parallels. It is about shifting our perspective to understand the ocean as a living system—one that responds, adapts, and evolves just as any organism would. The ocean's health is intimately tied to our own, and by seeing it as a living entity, we can better grasp the consequences of our actions. The pollutants we release, the carbon we emit, the ecosystems we overexploit—each of these assaults weakens the ocean's body, just as toxins or injuries weaken our own.

In *The Ocean's Body*, we will explore these connections in depth, drawing on the anatomy of the human body to illuminate the complex and often mysterious processes of the ocean. By seeing the ocean as a living organism—one that breathes, circulates, senses, defends, and heals—we can begin to appreciate its intricacies and vulnerabilities. More importantly, we can start to understand why it is essential to protect and nurture this vast, living system, because the ocean is not just something that surrounds us—it is part of us, as integral to life on Earth as our own bodies are to our individual survival.

This is not just a story of science; it is a call to rethink our relationship with the ocean. We must learn to care for it as we would care for our own bodies—because in many ways, the ocean *is* the body of the Earth, the force that sustains life, regulates climate, and connects all living things. Understanding this truth is the first step toward healing the damage we have done and ensuring that the ocean continues to thrive for generations to come.

CHAPTER 1

THE OCEANIC ORIGIN OF LIFE AND THE LIVING MEMORY OF SEAWATER

Long before the Earth was teeming with life—before the forests, deserts, and vast landscapes emerged—there was the ocean. In the ancient, primordial waters of early Earth, the first spark of life ignited. It was here, in this liquid world, that the earliest cells, the foundation of all life, began their journey. The ocean was not just a cradle of life; it was the origin itself—a place where chemistry met biology, and the building blocks of existence first came together.

Billions of years ago, the Earth's oceans were vastly different from what we know today. These seas were warmer, rich in minerals, and bathed in a cocktail of gases, such as methane, ammonia, and carbon dioxide, creating an environment ripe for the emergence of life. It was within this primordial soup that simple molecules began to organize themselves into more complex forms. The interactions of these molecules, catalyzed by the energy from underwater volcanic vents or perhaps lightning strikes, led to the formation of the first cell membranes—thin, protective layers that separated the inner workings of a primitive cell from the chaotic environment outside.

We do not know how the first biological molecules merged into living cells, but for sure these first cells were simple, likely resembling today's prokaryotes, organisms without a nucleus. Inside these microscopic bubbles, the ocean's waters were trapped, making them miniature versions of their environment. The ocean provided everything these early life forms needed—minerals, nutrients, and the stability to begin metabolic processes that would eventually lead to the first true living organisms. In this way, life was not just born in the ocean; it was a product of the ocean, with seawater forming the very first living environments.

The first cells in the primordial soup of the ocean. AI-generated image.

Cells: The Ocean's Legacy in Our Bodies

Today, we often think of the cells in our bodies as distinct, isolated units, working together in a complex organism. But in reality, every cell in our body carries with it the memory of the ocean's origin. Our cells, like those first ancient cells, are composed primarily of water—saltwater, to be specific. The fluids within our cells are strikingly similar to the composition of seawater (although at lower concentration), with sodium, potassium, chloride, and trace elements like calcium and magnesium. It is as if each of our cells is a tiny ocean, a living relic of our marine ancestry.

The way our cells function still mirrors the environment in which they evolved. Cell membranes, the boundary between the inside of the cell and the outside world, allow for the movement of ions and water, maintaining a delicate balance—just as the thermocline[1] maintain a balance between the upper and the lower layers of the ocean. Osmosis, the movement of water across these membranes, controls how our cells take in nutrients and expel waste, much like how the ocean's currents distribute nutrients and remove waste across vast distances. The mechanisms that keep our cells alive—those that regulate energy production, waste removal, and communication—are reflections of the processes that first evolved in the depths of the ocean.

Inside every human cell are tiny structures called mitochondria, which produce energy. These organelles have their own unique origin story, as they are believed to have evolved from ancient bacteria that once lived freely in the ocean. The early bacteria entered into symbiosis with larger cells, forming the complex, energy-producing structures that power all animal life today. This evolutionary partnership reminds us

---

[1] A thermocline is a layer in the ocean or a lake where the water temperature changes quickly with depth. Above it, the water is warmer, and below, it is much colder. This layer often forms during warmer months and affects marine life, currents, and nutrient mixing.

that even the energy production within our bodies is rooted in ancient marine life, a continuation of the ocean's creative force.

We Are Living Seawater

As humans, we often think of the ocean as a distant entity—something external that we visit, explore, or depend on for resources. But in reality, the ocean is not outside of us; when we sweat, we release salty water, an echo of the ancient marine environment our ancestors once lived in. When we cry, the salt in our tears reflects the same origin. Even the way our bodies regulate water and salt levels—through the kidneys and various cellular mechanisms—harkens back to the early evolutionary adaptations that helped organisms survive in the fluctuating salinity of the ocean.

The human body is essentially a vessel of living seawater, a system that depends on maintaining a precise balance of salts, minerals, and fluids—much like the ocean itself. Our bodies are constantly striving for homeostasis, the state of balance that allows life to thrive. This mirrors the ocean's own delicate equilibrium, where currents, temperature, and nutrient cycles must remain in balance to support the vast diversity of marine life. The processes that govern life in the ocean, from nutrient cycling to osmoregulation, are reflected within us, from our cellular function to our most complex biological systems.

In this way, we can understand that we are living seawater. Our bodies, our cells, our entire existence is a continuation of the ocean's original creation of life. Just as the ocean is vital for all living things, so too is it essential to recognize the deep connection we share with it—one that runs as deep as the tides and as old as the planet itself.

CHAPTER 2

THE SKIN OF THE SEA: THE OCEAN'S SURFACE AS A LIVING BOUNDARY

In many ways, the surface of the ocean is like the skin of a living organism—a boundary where the ocean interacts with the atmosphere, regulates temperature, breathes, and protects its deeper layers. The ocean's surface, or epipelagic zone, plays a crucial role in controlling many of the Earth's vital processes. Just as our skin shields and maintains our body's internal environment, the surface of the ocean serves as a dynamic, protective, and responsive layer that sustains life on the planet.

Unlike the human body, where skin is a solid, tangible barrier, the ocean's skin is a fluid boundary—a constantly shifting interface between the water and the air. It is here, in this thin upper layer, that some of the most important exchanges of gases, heat, and nutrients occur. It is also where life begins, with phytoplankton playing a critical role in regulating global oxygen levels and the carbon cycle. These processes make the ocean's surface more than just a boundary—it is a breathing, living interface that connects the sea and the sky.

The Physical Properties of the Ocean's Surface: A Breathing Boundary

Just as human skin regulates body temperature through sweating and blood flow, the surface of the ocean plays a key role in regulating the Earth's climate. The ocean's surface layer, which extends down to about 200 meters, absorbs vast amounts of solar energy. This heat is then distributed throughout the globe via ocean currents, influencing weather patterns, stabilizing temperatures, and driving the climate systems that sustain life on Earth. The ocean's skin acts as a buffer, absorbing and releasing heat depending on environmental conditions, much like the human body's skin adjusts to maintain homeostasis.

This layer is not static—it is constantly moving, mixing, and interacting with the air above it. Wind, waves, and temperature fluctuations create turbulence, which mixes the water and allows for the exchange of gases between the atmosphere and the ocean. Oxygen, carbon dioxide, and other gases are exchanged across this boundary, similar to how human skin allows for some gas exchange during respiration[2]. The surface of the ocean "breathes" in carbon dioxide from the atmosphere, storing it in the water, and "exhales" some (not much actually) of the oxygen produced by the photosynthetic activities of phytoplankton.

At the heart of this breathing process are phytoplankton. Similarly to the cells in human skin that provide protection and enable some gas exchange, phytoplankton play a vital role in regulating the composition of the atmosphere. Through photosynthesis, phytoplankton convert carbon dioxide into organic matter and oxygen, supplying nearly half

---

[2] Human skin plays a small but important role in the exchange of gases, allowing tiny amounts of oxygen to enter and carbon dioxide to leave. While it is not our primary method of breathing, the skin contributes to our overall health by supporting this subtle process, along with protecting the body and regulating temperature. This gentle form of respiration helps maintain balance in our body's systems.

of the Earth's oxygen[3]. When phytoplankton die or are consumed by other organisms, some of this organic carbon sinks to the ocean floor, where it can remain sequestered for thousands of years. This process, called the biological pump (see Chapter 4), effectively removes carbon from the atmosphere, helping to regulate the Earth's climate and prevent excessive global warming. Without phytoplankton, atmospheric CO_2 levels would rise significantly, leading to more extreme climate changes and threatening the stability of marine and terrestrial ecosystems alike.

Phytoplankton and Ocean Color: Regulating Light Absorption

Much like the skin's melanin controls how much sunlight the body absorbs, phytoplankton influence how much sunlight is absorbed or reflected by the ocean's surface. When phytoplankton populations are high, they change the color of the water, increasing its albedo—the ability to reflect sunlight. During large phytoplankton blooms, the ocean's surface appears greener, which can reflect more sunlight back into space and reduce the amount of heat absorbed by the ocean.

This ability to regulate sunlight absorption helps control ocean temperatures and prevents excessive warming of surface waters. In this way, phytoplankton play a role in the Earth's heat budget, helping to cool the planet in a manner similar to how human skin prevents overheating by releasing sweat to cool the body.

The presence of phytoplankton also enhances the ocean's role as a carbon sink. By capturing sunlight and converting carbon dioxide into organic carbon, phytoplankton reduce the amount of solar radiation

---

[3] While it is true that phytoplankton produces the same (ore even more) amount of oxygen tan terrestrial plants, most of it is consumed by the marine organisms; only a small fraction reaches the atmosphere in very productive systems.

absorbed by the water, while simultaneously sequestering carbon that would otherwise contribute to the greenhouse effect. This dual function—reducing heat absorption and storing carbon—helps to moderate global temperatures.

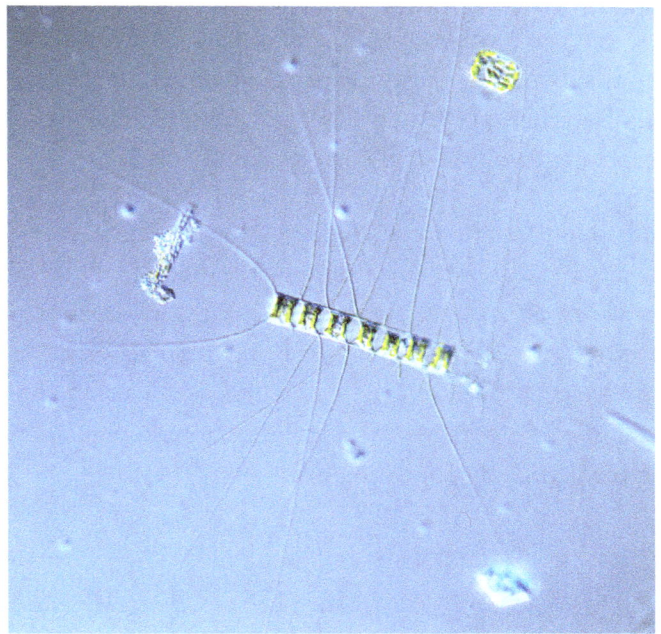

Marine diatom. Phytoplankton

The Ocean's Skin in Crisis: Protecting the Boundary Between Earth and Sky

In the same way that human skin is vulnerable to environmental damage, the surface of the ocean is under threat from human activities. The warming of ocean waters, driven by climate change, is altering the delicate balance of the ocean's skin. This warming not only affects the health of phytoplankton but also reduces the ocean's ability to absorb carbon dioxide, further exacerbating global warming. The ocean's surface, once a powerful buffer for atmospheric carbon, is becoming less effective as its physical and chemical properties are disrupted.

Plastic pollution, classic pollutants (heavy metals, hydrocarbons,) and new emergent pollutants (plasticizers, flame retardants, medicines, hormones, etc.) are other major threat to the ocean's skin. These particles and substances disrupt the biochemical processes that occur at the ocean's surface and enter plankton and other organisms, leading to toxic buildups throughout the food chain.

The ocean's skin is not just a passive layer; it is an active, living interface that plays a critical role in maintaining the health of the planet. To protect it, we must address the root causes of its decline. Reducing carbon emissions, managing pollution, and protecting marine ecosystems are essential steps in healing the ocean's skin and ensuring that it continues to perform its vital functions—regulating climate, producing oxygen, and supporting life.

CHAPTER 3

THE CIRCULATORY SYSTEM: OCEAN CURRENTS AS BLOOD VESSELS

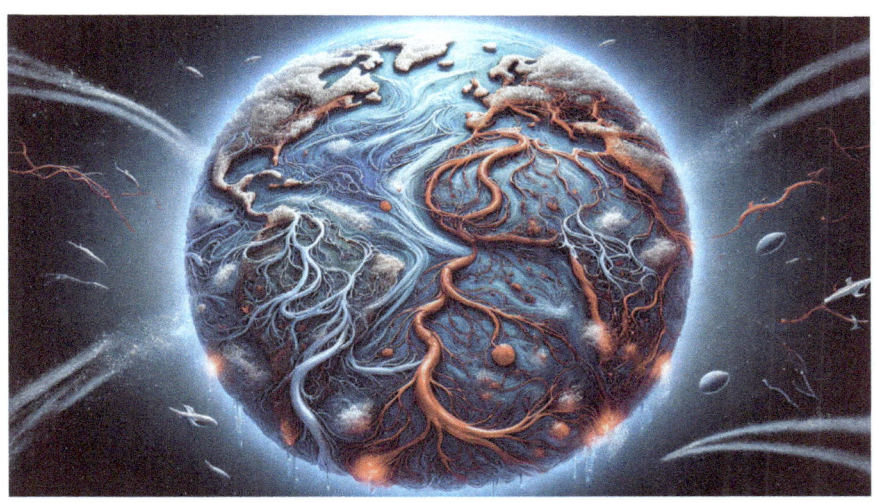

In the human body, the circulatory system acts as the lifeline that transports oxygen, nutrients, and waste products to and from cells, ensuring that every organ and tissue receives the resources necessary to function properly. The ocean, too, possesses a circulatory system—a vast and intricate network of currents that move nutrients, oxygen, and marine life throughout its expansive waters. These ocean currents, driven by wind, the Earth's rotation, and differences in temperature and salinity, serve as the "blood vessels" of the ocean, circulating vital resources and playing a crucial role in the health of the planet.

Ocean circulation is essential not only for the distribution of nutrients and organisms but also for regulating the Earth's climate and supporting ecosystems. This chapter will delve into the ways in which ocean currents perform functions akin to those of the human circulatory system, with a focus on nutrient distribution, heat transport, and ecosystem maintenance. By examining different types of ocean currents—such as surface currents, deep-water circulation, and upwelling zones—we can better understand how these forces drive the movement of marine life and influence global climate patterns.

Surface Currents: The Arteries and Veins of the Ocean

In the human body, arteries transport oxygen-rich blood from the heart to the rest of the body, while veins return deoxygenated blood back to the heart. A similar pattern exists in the ocean, where surface currents carry warm, nutrient-poor water from the equator toward the poles and return cold, nutrient-rich water from polar regions toward the equator. This exchange plays a critical role in regulating global climate, redistributing heat, and sustaining marine ecosystems.

Surface currents, driven primarily by wind and the Coriolis effect[4] (which deflects moving fluids due to the Earth's rotation), form large-scale circulation systems known as gyres. These gyres function like the body's circulatory loops, moving water and distributing heat, organisms, and nutrients across ocean basins in a circular pattern, much like blood vessels in the human body circulate blood through major arteries. The Earth's five major gyres—the North Atlantic, South Atlantic, North Pacific, South Pacific, and Indian Ocean gyres—are the ocean's primary circulatory loops, driven by trade winds and westerlies.

Gyres are critical for maintaining the balance of nutrients in the ocean. In regions where gyres intersect with upwelling zones, they bring nutrient-rich waters to the surface, fueling the productivity of entire ecosystems. This process is comparable to how blood vessels ensure the distribution of oxygen and nutrients to different parts of the body.

---

[4] The Coriolis effect is the apparent deflection of moving objects, such as air or water, caused by the Earth's rotation. In the Northern Hemisphere, it causes objects to veer to the right, and in the Southern Hemisphere, to the left.

Upwelling: The Capillaries of the Ocean

In the human body, capillaries surrounding lung alveoli are responsible for delivering oxygen into the bloodstream, facilitating gas exchange and ensuring that the body receives the oxygen it needs to survive. Similarly, upwelling zones[5] in the ocean act as the "alveoli" of the marine environment, bringing nutrient-rich deep waters to the surface, where they "oxygenate" marine ecosystems by fueling the growth of phytoplankton. These phytoplankton, much like the oxygen carried by blood, are the lifeblood of the ocean, forming the base of the marine food web and nourishing entire ecosystems.

Coastal upwelling zones, such as those along the coasts of Peru, California, and Namibia, can be thought of as the ocean's most productive "lungs." Just as the alveoli efficiently transfer oxygen into the bloodstream, these regions draw up essential nutrients like nitrogen and phosphorus, supporting massive blooms of phytoplankton. These blooms, in turn, sustain vast populations of zooplankton, small fish, and predators such as seabirds and marine mammals. In both systems, the smallest units—alveoli in the lungs and upwelling zones in the ocean—are vital for maintaining overall health. Without the continuous flow of oxygen from lung capillaries or nutrients from upwelling, the human body and the marine ecosystem would suffer from a lack of critical resources.

Equatorial upwelling, where trade winds cause surface waters to diverge along the equator, functions similarly to capillaries returning oxygenated blood to the heart. It replenishes nutrient-depleted surface waters with compounds from the depths, ensuring that life-sustaining nutrients are circulated where they are most needed. Without these

---

[5] Upwelling is the process in which deep, cold, and nutrient-rich water rises to the surface, replacing warmer surface waters. This occurs mainly along coastlines due to wind patterns that push surface water away from the shore, allowing deeper water to rise.

oceanic "alveoli," marine life would be starved of essential resources, much as the body would be deprived of oxygen without functional lung capillaries.

Plankton as the blood cells of the ocean

In the human body, red blood cells are responsible for transporting oxygen to tissues, sustaining the body's life processes. Similarly, phytoplankton serve as the primary producers in the ocean, generating oxygen through photosynthesis and providing the foundation of the marine food web. Much like red blood cells deliver oxygen to all parts of the body, phytoplankton fuel marine ecosystems by producing the energy and nutrients that support a vast array of marine organisms. Zooplankton, which feed on phytoplankton, can be compared to white blood cells in their role as regulators of the ecosystem. Akin white blood cells defend the body against pathogens, zooplankton manage the balance of plankton populations by grazing on phytoplankton and transferring energy to higher trophic levels. This ensures the proper functioning of the ocean's biological "immune system," keeping ecosystems in balance and supporting biodiversity. Additionally, planktonic larvae and eggs, which drift with the currents to new environments, can be likened to stem cells in the human body. Stem cells have the unique ability to develop into specialized cells, much like larvae and eggs grow into diverse marine species. These drifting life stages, carried by ocean currents, represent the ocean's potential for renewal and regeneration, ensuring the resilience and continued evolution of marine populations. In this way, the various components of plankton function similarly to blood cells, each playing a vital role in maintaining the health, balance, and sustainability of the ocean's vast ecosystems.

In addition to plankton, many marine organisms, such as fish, mollusks, and crustaceans, rely on currents to disperse their larvae.

Larval dispersal is crucial for maintaining genetic diversity and allowing species to colonize new habitats. Just as the circulatory system helps distribute oxygen and nutrients throughout the human body, ocean currents distribute larvae across vast distances, enhancing the survival of species and promoting ecosystem resilience. Without this natural dispersal mechanism, marine populations could become isolated, reducing biodiversity and making ecosystems more vulnerable to environmental changes.

Thermohaline Circulation: The Heart and Major Blood Vessels of the Ocean

While surface currents circulate water across the upper layers of the ocean, the deep ocean is dominated by thermohaline circulation—a system of currents driven by temperature and salinity differences. This global conveyor belt begins with the sinking of cold, dense water in polar regions, functioning like the deep circulatory pathways of the human body. It ensures that nutrients and oxygen reach even the most remote parts of the ocean, similar to how blood vessels distribute oxygen and nutrients throughout the body.

Deep water formation in regions such as the North Atlantic can be compared to how blood returns to the heart after being depleted of oxygen. Cold, dense water sinks into the deep ocean, beginning a slow journey around the globe. This movement is vital for sequestering carbon and redistributing nutrients, much like the circulatory system nourishes the body. As this deep water travels, it gathers organic matter from decaying organisms, enriching itself with nutrients that will eventually resurface in upwelling zones. The continuous flow of this conveyor belt is crucial for maintaining the health of the ocean, just as the circulatory system is essential for the health of the body. Without this movement, the ocean would stagnate, cutting off essential resources and potentially leading to ecosystem collapse.

At the surface, the Gulf Stream, often compared to the heart of the ocean, drives one of the most powerful currents in the Atlantic. Like the heart pumping blood through arteries, the Gulf Stream transports warm water from the tropics to northern latitudes. This current has a profound impact on weather patterns and marine life, influencing everything from hurricanes to fish migration. It acts as the ocean's pulse, delivering warmth and life to regions far removed from the equator, much as the heart sustains organs by keeping blood in motion.

AI-generated image representing the similarity between human circulatory system and ocean currents

The Impact of Climate Change on Ocean Circulation

Just as the human circulatory system can be disrupted by blockages or disease, the ocean's circulation is vulnerable to disruption from climate change. Rising global temperatures, changing wind patterns, and the melting of polar ice are all affecting the delicate balance of ocean currents, not only the Gulf Stream, with potentially severe consequences for marine ecosystems.

As the surface layers of the ocean warm, increased stratification (the separation of surface and deep waters) reduces the mixing of nutrient-rich deep water with nutrient-depleted surface layers. This stratification is akin to poor circulation in the human body, where restricted blood flow can lead to tissue damage and poor health. In the ocean, reduced nutrient mixing can limit phytoplankton growth, diminishing the base of the marine food web and threatening the productivity of ecosystems that depend on upwelling.

In polar regions, the melting of ice caps is diluting the salinity of surface waters, weakening the process of deep water formation that drives thermohaline circulation. If this circulation weakens or halts, the global conveyor belt could slow dramatically, cutting off the flow of nutrients and oxygen to the deep ocean. This disruption can be likened to heart failure in the human body, where the failure of the heart to pump blood effectively leads to systemic collapse. If the global conveyor belt falters, marine ecosystems could face catastrophic consequences, with cascading effects on climate regulation, food webs, and biodiversity.

CHAPTER 4

THE DIGESTIVE SYSTEM: ZOOPLANKTON AS THE GUT OF THE OCEAN

The human digestive system breaks down food into nutrients that cells can use for energy, growth, and repair. Similarly, the ocean has its own digestive system, a complex network of organisms and processes that break down organic matter and recycle nutrients, supporting the productivity of marine ecosystems. Central to this system are zooplankton, the primary consumers of the ocean, which act as the "gut" by feeding on phytoplankton and transferring energy up the food chain. Zooplankton play a role analogous to the human digestive system by converting raw material into usable energy for marine organisms, while also recycling nutrients back into the ecosystem. However, another crucial component of this digestive system is the ocean's bacterial communities, which function much like the microbiota in the human gut.

The Mouth: Ocean's Surface and Coastal Zones

Just as the human mouth is the entry point for food, the ocean's surface and coastal zones are where nutrients first enter the marine system. In the human body, enzymes in saliva start breaking down carbohydrates as soon as food enters the mouth. Similarly, at the ocean's surface and coastal zones, primary producers like phytoplankton immediately begin to utilize the nutrients that enter from rivers, rainfall, or atmospheric deposition. These nutrients, combined with sunlight, allow phytoplankton to photosynthesize, transforming inorganic substances into organic matter that forms the foundation of marine food webs.

This matter then becomes available for zooplankton, the primary consumers. The surface layers of the ocean, where most of this primary production occurs, can be also thought of as other "mouths" of the ocean, where raw materials are processed and prepared for the next steps in the digestive journey.

Zooplankton: The Ocean's Digestive System

Zooplankton are the vital "stomach" of the ocean, processing the organic matter produced by phytoplankton and converting it into forms usable by higher organisms. Zooplankton range from microscopic protozoans to larger crustaceans like copepods and krill, and even include gelatinous creatures such as jellyfish. Just as the human stomach breaks down food into simpler components that the body can absorb, zooplankton consume phytoplankton and break down the energy stored in these tiny organisms.

When zooplankton feed on phytoplankton, they convert this energy into biomass, which is then transferred up the food chain to larger marine predators such as fish, whales, and seabirds. This process is akin to the small intestine in humans, where nutrients from digested food are absorbed and distributed throughout the body to fuel various functions. Zooplankton, in their central role, ensure that the energy produced by phytoplankton is accessible to the rest of the marine ecosystem, acting as a critical conduit between the base of the food web and higher trophic levels.

Without zooplankton, the energy generated through photosynthesis would be locked at the base of the food chain, mostly inaccessible to larger organisms. This would lead to a collapse of marine food webs, much like how a malfunctioning human stomach would prevent the body from absorbing the nutrients needed to survive. The intricate feeding habits of zooplankton, whether filter-feeding on small particles or actively hunting other microscopic prey, mirror the versatility of the human digestive system in breaking down diverse food sources.

Marine copepod

The Role of Bacteria: The Ocean's Microbiome

In the human digestive system, beneficial bacteria play an essential role in breaking down complex carbohydrates, producing vitamins, and aiding in the absorption of nutrients. Similarly, in the ocean, bacterial communities perform vital functions in nutrient cycling and organic matter decomposition, acting much like the microbiota in the human gut.

Bacteria in the ocean contribute to what is known as the microbial loop, where they break down dissolved organic matter (DOM) and detritus—organic debris that results from the decomposition of dead organisms or waste products. This breakdown process releases nutrients back into the water, making them available for primary producers like phytoplankton. In this way, bacteria function much like gut bacteria that break down indigestible food and release nutrients that can be absorbed by the body.

Additionally, ocean bacteria play a crucial role in nitrogen fixation, converting nitrogen gas from the atmosphere into forms that marine organisms can use. This is comparable to how gut bacteria produce certain vitamins, such as vitamin K, which the human body cannot synthesize on its own. Without these bacterial processes, essential nutrients like nitrogen and phosphorus would not be recycled effectively, reducing the ocean's productivity and disrupting marine food webs.

Bacteria are also involved in the breakdown of organic matter on a larger scale, particularly in the deep ocean and sediments, where they decompose the remains of dead organisms. This process is critical for nutrient regeneration and plays a central role in maintaining the balance of marine ecosystems, similar to how gut bacteria help maintain the overall health of the human digestive system.

Nutrient Cycling and the Biological Pump

Nutrient cycling in the ocean mirrors the human body's method of reusing and redistributing essential compounds. As zooplankton feed on phytoplankton, they excrete waste that releases nutrients such as nitrogen and phosphorus back into the water. These nutrients are critical for the growth of primary producers, creating a continuous cycle that sustains the productivity of the ocean, much like how the human body conserves and reallocates nutrients to support vital functions.

Ocean bacteria play a crucial role in this cycle by breaking down the waste products produced by zooplankton and other marine organisms. This microbial activity converts organic matter into simpler compounds that can be absorbed by phytoplankton, allowing the nutrient cycle to continue. Without this bacterial involvement, much of the ocean's organic material would remain unprocessed, limiting the availability of essential nutrients and decreasing the overall productivity of the marine ecosystem.

In addition to nutrient recycling, zooplankton contribute to the biological pump, a process by which carbon, in the form of organic material, is transferred from the surface ocean to the deep sea. As zooplankton consume phytoplankton, they incorporate carbon into their bodies and release some of it through waste products. When these organisms die or produce waste, a portion of this carbon sinks to the ocean floor, where it can be stored for centuries or even millennia. This process of carbon sequestration helps regulate atmospheric CO_2 levels and mitigate climate change, functioning much like how the human body stores excess energy in fat reserves for future use.

Zooplankton as Prey: Fueling Marine Ecosystems

Zooplankton not only recycle nutrients and facilitate energy transfer, but they also serve as a vital food source for a wide variety of marine predators. For example, krill are the primary prey for baleen whales, which filter vast quantities of these tiny organisms from the water. In this role, zooplankton are like the stomach's function in providing nutrients to other parts of the body, converting primary production into energy that sustains larger marine animals.

Small fish, such as herring, sardines, and anchovies, feed extensively on zooplankton, and these fish, in turn, are prey for larger predators such as sharks, seabirds, and marine mammals. This energy flow supports the entire marine ecosystem, creating a chain of life that sustains both small and large organisms alike. Without zooplankton, the productivity at the base of the food web would be inaccessible to higher trophic levels, leaving larger animals without the energy they need to survive. In the human body, this would be akin to a breakdown in the digestive system that prevents nutrients from reaching the organs and tissues that rely on them.

Threats to the Ocean's Digestive System

The ocean's delicate digestive system, driven by zooplankton and bacteria, faces significant threats from human activities and climate change. Rising ocean temperatures disrupt the timing of phytoplankton blooms, which zooplankton rely on for food. Warmer waters also affect the distribution of zooplankton species, especially in polar and temperate regions, where these shifts can have far-reaching consequences. These changes are akin to digestive disorders in humans, where the body's ability to process and distribute nutrients becomes impaired, leading to broader health problems.

Ocean acidification, caused by increasing levels of atmospheric CO_2, poses a specific threat to species that rely on calcium carbonate to build their shells, such as coccolithophores, foraminifera, and pteropods. Acidification weakens these shells, making these organisms more vulnerable to predation and reducing their ability to survive and reproduce. This weakening of the ocean's "gut" is comparable to a malfunctioning digestive system in humans, where the breakdown and absorption of nutrients become less efficient, leading to a decline in overall health. Unfortunately, there is no simple solution or "antacid" to alleviate the stress we are placing on the ocean's systems.

The ocean's digestive system, much like our own, is a complex, interconnected web of processes that sustains life. Protecting this system is essential not only for marine ecosystems but for the health of the planet as a whole.

CHAPTER 5

THE IMMUNE SYSTEM: DEFENSES AND TOXINS

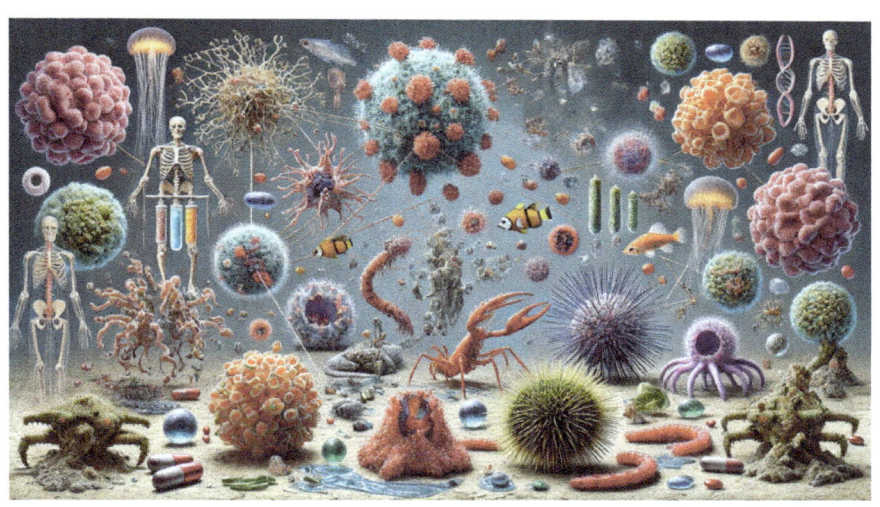

Our immune system functions as a coordinated defense mechanism, integrating various specialized cells and processes to maintain health and balance. Although these responses seem distinct in their roles—barriers, chemicals, and immune cells—they operate as a system that defends against external threats and maintains homeostasis. Similarly, the ocean can be viewed as a living body with its own "immune system," where a diverse range of organisms and ecosystems collaborate to protect against stressors, such as predation, environmental changes, and human impacts.

From the smallest plankton to the largest marine predators, each species plays a role in this vast network of defense mechanisms, ensuring the stability and resilience of the ocean as a whole. While individual species may employ unique strategies—like shells, toxins, or evasive behaviors—these responses are part of an interconnected system. Together, they contribute to the ocean's ability to maintain ecological balance, adapt to changes, and recover from disturbances.

In this chapter, we will not only explore these individual strategies but also examine how they integrate into a larger "immune" network, mirroring the collective defenses of the human body. Just as the immune system's various components interact to protect against disease, the organisms of the ocean work together, forming a complex web of protection that helps the ocean thrive despite continuous challenges.

Physical Defenses: Armors and Barriers in the Marine World

Just as the human immune system relies on physical barriers like skin and mucous membranes to protect the body from infections and injuries, many marine organisms have developed physical defenses to shield themselves from predators and environmental threats. These defenses can take various forms, such as hard exoskeletons, shells, spines, and camouflage, serving as protective mechanisms against the many dangers present in the marine environment.

Marine species, such as mollusks, crustaceans, and certain types of plankton, have evolved hard exoskeletons or shells that serve as their primary protection. Bivalves like clams and oysters use their calcium carbonate shells to defend against predators, while crustaceans like crabs and lobsters rely on their thick exoskeletons for physical protection. Even coccolithophores, a type of phytoplankton, construct intricate calcium carbonate shells, called coccoliths, which help shield them from predators and environmental changes. These hard protective structures function similarly to the human skin and mucous membranes, providing a crucial first line of defense against external threats.

Other marine creatures, such as sea urchins, rely on physical armor like spines to fend off predators. These spines create an effective deterrent, discouraging fish and other animals from attempting to eat them. Similarly, pufferfish have developed the ability to inflate their bodies, revealing sharp spines that make them less attractive to potential predators. These forms of physical defense are essential for marine organisms, particularly those at the base of the food web, such as plankton, which are often unable to escape predators through mobility. By developing structures like hard shells and spines, these organisms enhance their chances of survival in an environment where predation is a constant and intense threat.

Chemical Defenses: Toxins and Bioluminescence

In addition to physical barriers, many marine species have evolved chemical defenses to protect themselves from predators. This mirrors how the human immune system uses antibodies and other chemicals to neutralize harmful pathogens. The chemical defenses employed by marine organisms can include toxic compounds that deter predators or bioluminescent displays that confuse or frighten potential threats.

Certain marine species produce toxins or venoms to defend themselves. For example, the venomous tentacles of box jellyfish can deliver a powerful sting to anything that comes into contact with them. Cone snails, on the other hand, produce highly potent venom, which they use to immobilize both prey and predators. Some algae and plankton, such as dinoflagellates, produce toxins that can lead to harmful algal blooms (HABs), commonly known as "red tides." These blooms deter predators and can have harmful effects on marine life and even humans. These toxic compounds operate much like antibodies in the human immune system, neutralizing threats before they can cause harm.

Bioluminescence is another fascinating chemical defense mechanism found in marine life. Many organisms, from deep-sea fish to plankton such as dinoflagellates, use bioluminescence as a means of protection. When disturbed, bioluminescent plankton emit light, creating glowing clouds in the water that can startle or confuse predators, giving the plankton a chance to escape. Other species, such as the anglerfish, use bioluminescence both as a defensive tool and as a means to lure prey. In some cases, bioluminescence is used for camouflage, allowing organisms to blend in with the faint light from the surface, making them nearly invisible to predators in the dark depths of the ocean. This behavior is reminiscent of how certain immune cells in the human body can mask pathogens or infected cells, preventing them from being detected by the immune system.

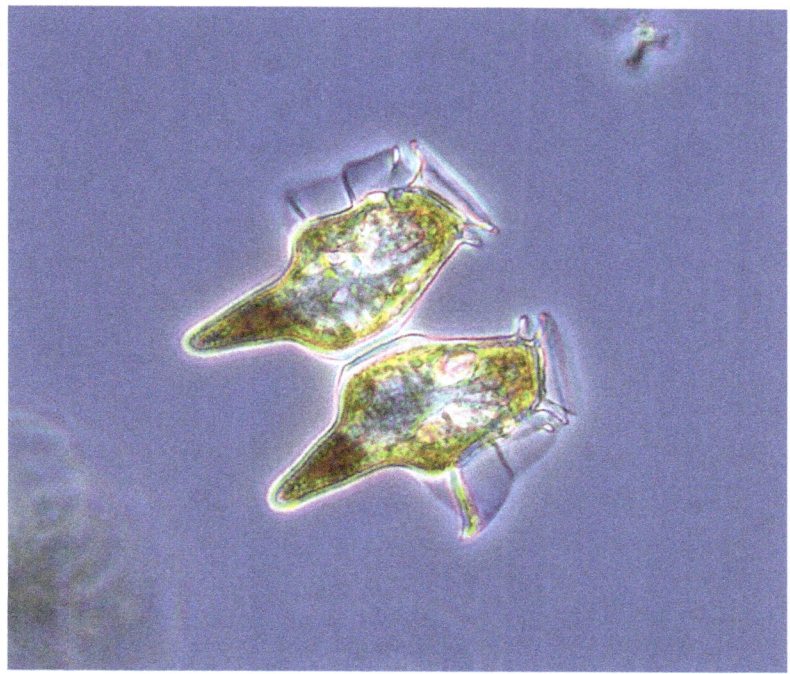

Toxic dinoflagellate producing HABs

Behavioral Defenses: Evasion, Mimicry, and Camouflage

Marine organisms also rely on behavioral adaptations to protect themselves from threats, just as the human body employs certain behaviors—like fever or coughing—to combat infections. These behavioral defenses enable marine organisms to evade predators, escape environmental stressors, and increase their likelihood of survival in a hostile world.

Many marine species rely on speed and mobility to escape from predators. Fish often swim in schools, creating a confusing mass that makes it difficult for predators to single out an individual. Flying fish have developed the ability to leap out of the water and glide through the air for short distances, allowing them to evade predators like tuna and dolphins. Zooplankton, such as copepods, use a different strategy: they perform daily vertical migrations, staying in deeper, darker waters during the day to avoid visual predators, and rising to the surface at night to feed on phytoplankton. This behavior allows them to reduce the risk of predation, similar to how the human immune system regulates inflammation depending on the level of threat in different areas of the body.

Mimicry and camouflage are additional behavioral adaptations that many marine organisms use to avoid detection. Octopuses and cuttlefish are well-known for their mastery of camouflage, altering their color, texture, and shape to blend into their surroundings. This remarkable ability allows them to avoid predators and ambush prey with ease. Flatfish also use camouflage, changing their skin color to match the ocean floor, effectively becoming invisible to both predators and prey. Mimicry is another powerful strategy used by some species, like the mimic octopus, which can imitate the appearance and behavior of more dangerous animals, such as venomous sea snakes, to deter potential threats. This mimicry is akin to how the human immune system identifies patterns in pathogens and tailors its response to specific threats, ensuring that the body can adapt to different types of dangers.

Environmental Defenses: Coping with Stressors and Change

Just as the human immune system can be overwhelmed by chronic stress, disease, or injury, the ocean's natural defenses are increasingly tested by human activities. Overfishing, habitat destruction, climate change, and pollution are placing immense pressure on marine species, threatening their ability to maintain the balance of the ocean's "immune system."

Overfishing, for example, disrupts the balance of marine ecosystems by removing key species, particularly predators, from the food web. This leads to trophic cascades, where the loss of predators causes an overabundance of prey species, which in turn depletes their food sources. The result is a degraded and unbalanced ecosystem, similar to how an overactive or weakened immune system can lead to autoimmune diseases or chronic infections.

Another visible sign of the ocean's struggle to cope with environmental stressors is coral bleaching, caused by rising sea temperatures and ocean acidification. When the water becomes too warm, corals expel the symbiotic algae that live within their tissues, losing their main source of energy. Without these algae, corals become more susceptible to disease and death, affecting the entire ecosystem that depends on coral reefs for shelter and food. Coral bleaching is akin to a breakdown in the human immune system, where the body's defenses are overwhelmed, and the health of the entire organism is compromised. As climate change accelerates, the ocean's ability to defend itself against these stressors will be increasingly tested.

Fortunately, in many instances just like as the human immune system responds to external stressors such as pathogens, toxins, and allergens, marine organisms can cope with changing environmental conditions. For instance, certain types of coral can acclimatize to higher temperatures by forming symbiotic relationships with heat-resistant strains of algae. This adaptation allows them to survive in conditions that would otherwise lead to coral bleaching and death. Other marine organisms, such as fish and crustaceans, have developed mechanisms of thermal tolerance, enabling them to survive within a wider range of temperatures. This ability is similar to how the human immune system can adapt to different pathogens and toxins, producing new antibodies to fight infections and respond to environmental stress.

Some species have evolved the ability to tolerate or detoxify pollutants. For example, certain bacteria can break down toxic compounds like oil or pesticides into less harmful substances. Similarly, some species of mollusks can filter pollutants from the water, storing them in their tissues without suffering harm. These detoxifying abilities are comparable to the function of the human liver, which processes and neutralizes toxins in the bloodstream.

Moreover, marine ecosystems, much like the immune system, rely on natural processes and species to restore balance. Mangroves and seagrass meadows act as nature's filters, absorbing excess nutrients and pollutants, thereby preventing eutrophication. For instance, seagrasses trap sediments, improving water clarity and reducing harmful algal blooms. Similarly, kelp forests help sequester carbon and act as buffers against coastal erosion.

Keystone species like sea otters play a crucial role in controlling prey populations, such as sea urchins, which would otherwise overgraze kelp forests. Without otters, the unchecked urchin population can decimate kelp, similar to how the absence of immune cells allows infections to spread unchecked. Sharks, another keystone species, regulate fish populations, maintaining the balance of the food web and preventing the overpopulation of certain species that could destabilize ecosystems.

Natural ecosystem recovery can also be seen through resilience mechanisms. For example, after a disturbance like a hurricane, coral reefs can recover if nearby fish populations (like parrotfish) keep algae in check, allowing corals to regrow. Seabirds and fish migrations transport nutrients across ecosystems, redistributing essential elements that facilitate recovery.

These strategies demonstrate that, much like the human immune system, ecosystems possess built-in mechanisms to cope with and recover from stressors. However, just as the human immune system's capacity to adapt has limits, marine ecosystems can only cope with stressors up to a point. When stressors like pollution, habitat destruction, or climate change overwhelm these natural defenses, ecosystems can collapse. These natural adaptations and recovery strategies underscore the need to mitigate environmental damage before tipping points are reached.

CHAPTER 6

THE BRAIN AND NERVOUS SYSTEM OF THE OCEAN

In the human body, the brain serves as the control center, regulating bodily functions, responding to stimuli, and maintaining balance across various systems. Similarly, the ocean operates like a living organism, where marine life collectively acts as its "brain" and "nervous system," regulating and shaping its physical, chemical, and biological dynamics. Unlike the centralized structure of the human brain, the "brain of the sea" is decentralized, formed by a complex network of organisms that function as ecological engineers, influencing the health of the ocean and maintaining its balance. These organisms—from plankton to marine mammals, coral reefs, fish, and benthic species—act together to sustain the ocean's equilibrium, foster biodiversity, and even regulate the planet's climate. In this way, marine ecosystems adapt and respond to both local and global changes, from nutrient availability to climate fluctuations. This distributed model aligns more with the concept of swarm intelligence found in certain animal species, where collective behaviors result in coordinated responses without centralized control, much like how the peripheral nervous system responds to local stimuli without needing direct input from the brain for every action.

This chapter will explore, then, how marine organisms function both as the "brain" and "nervous system" of the ocean, playing crucial roles in maintaining ecological processes, responding to environmental shifts, and regulating vast ecosystems. We will examine the vital contributions of organisms such as coral reefs, whales, dolphins, and schooling fish, understanding how their behaviors and interactions mirror the functions of the human brain and nervous system. Additionally, we will explore the growing human-induced challenges—such as pollution, climate change, and overfishing—that threaten the ocean's ability to maintain its balance.

Coral Reefs: Sensory Networks and the Memory of the Ocean

Just as the human brain processes sensory information and stores memories, coral reefs serve as critical nodes within the ocean's ecological network, acting as sensory hubs that maintain biodiversity, store genetic information, and facilitate communication between species. Coral reefs are often likened to the rainforests of the sea because of their immense biological diversity and their role in housing countless species of fish, invertebrates, and marine mammals. Much like the neurons in the brain transmit signals and coordinate functions, coral reefs help shape the surrounding environment by modifying water flow, stabilizing sediments, and providing essential ecosystem services that enable marine species to thrive.

Coral reefs actively engineer their environment through their calcium carbonate skeletons, building vast structures that influence local hydrodynamics. These reef structures create sheltered areas that slow down water currents, protect coastlines from erosion, and regulate nutrient cycling within the system. This ecological engineering mirrors the brain's ability to influence the body's internal balance, facilitating the proper functioning of all systems. Coral reefs are essential for storing genetic diversity (evolutive knowledge), acting as reservoirs of biological memory that help ecosystems withstand and recover from environmental stressors. Their resilience ensures the long-term health of marine ecosystems, much like the brain's capacity to learn and adapt through neural plasticity.

Coral reefs also play a vital role in sensing environmental changes. Corals are highly sensitive to fluctuations in temperature, acidity, and light. When conditions deteriorate, as seen with rising sea temperatures, coral bleaching occurs, signaling a breakdown in the reef's capacity to maintain equilibrium. This bleaching is similar to how the human brain reacts to stress, activating defense mechanisms, harmful in excess (such as cortisol), to preserve its overall function. Coral reefs, therefore, act as early warning systems, much like the sensory organs that help the brain detect and respond to external threats.

Coral reef, Australia

The Deep Sea: The Ocean's Subconscious and Memory Storage

The deep sea, with its cold, dark, and high-pressure environment, can be likened to the subconscious mind of the ocean, where processes occur out of sight yet play a critical role in the long-term health of marine ecosystems. Deep sea organisms such as benthic creatures, deep sea fish, and chemosynthetic bacteria perform vital functions in carbon sequestration, nutrient recycling, and the storage of organic material, ensuring the stability of the ocean's biogeochemical cycles.

One of the most important functions of the deep sea is its role in carbon sequestration. Organic material produced at the surface, including dead plankton and marine snow, sinks to the deep ocean, where it is broken down by bacteria and other microorganisms. This process removes carbon from surface waters and stores it in deep-sea sediments, mitigating climate change by regulating atmospheric CO_2 levels. The deep sea's ability to store carbon is similar to the brain's ability to store memories and information for long periods of time. Just as the brain consolidates information during rest, the deep sea acts as a repository for carbon and organic matter, maintaining the long-term stability of the planet's climate system.

Marine Mammals: Cognitive Control and Long-Distance Communication

Marine mammals—especially whales and dolphins—are often referred to as the ocean's "intelligent" creatures because of their complex social structures, cognitive abilities, and advanced communication systems. In the ocean's metaphorical brain, these mammals perform functions akin to higher-level cognition, influencing ocean dynamics, regulating predator-prey relationships, and impacting nutrient distribution through their movements and behaviors. Their actions have profound ecological consequences, shaping marine environments much like the brain governs complex behaviors and physiological functions.

Whales, particularly baleen whales, play an essential role as ecological engineers by participating in nutrient cycling through a process known as the whale pump. As they dive to feed in deeper waters and return to the surface to breathe, they release nutrients such as nitrogen and iron through their waste. These nutrients stimulate the growth of phytoplankton, the primary producers at the base of the marine food web. In this sense, whales act as "connectors," linking deep and surface ecosystems, facilitating the movement of nutrients across vast distances, and promoting primary production that supports entire marine ecosystems. Their movements, akin to the brain's neurons transmitting signals, help maintain connectivity in marine environments, ensuring that ecosystems remain productive and interconnected.

Dolphins, known for their advanced social structures and communication systems, engage in behaviors that resemble the brain's ability to coordinate complex tasks. Through their vocalizations, echolocation, and body language, dolphins communicate over long distances, enabling them to maintain group cohesion, coordinate hunting, and navigate the vast ocean. The ability of dolphins to "talk" across the sea mirrors the brain's capacity to transmit information through neural pathways. These networks of communication help regulate predator-prey dynamics, maintain balance in the food web, and ensure the health of fish populations, demonstrating how marine mammals serve as cognitive controllers within the ocean's metaphorical brain.

Schooling Fish: Reflexes and Collective Intelligence

In the human body, reflexes are automatic responses to external stimuli, enabling quick reactions to danger. In the ocean, schooling fish exhibit a form of collective intelligence that allows them to respond swiftly and cohesively to predators or changes in their environment. This behavior is comparable to the brain's reflexive control over the body's responses, ensuring immediate reactions to preserve survival.

Fish that swim in schools benefit from the confusion they create for predators, making it difficult for predators to single out individual fish. This synchronized movement is akin to how the human brain coordinates reflexes, allowing for rapid responses to potential threats. The collective behavior of schooling fish is a form of distributed decision-making, where each fish responds to changes in the group's movement almost instantaneously. This self-organized behavior mirrors the brain's autonomic nervous system, which controls involuntary functions like heart rate and digestion.

Fish school

Schooling fish also play a crucial ecological role in transferring energy and nutrients across marine ecosystems. As they migrate, feed, and spawn in large numbers, species such as herring, sardines, and anchovies link coastal ecosystems with deeper waters, transporting nutrients that support various marine predators. Their migrations and behaviors help maintain the flow of energy through the food web, much like how the brain regulates the distribution of resources throughout the body. The reflexive and collective intelligence of schooling fish is vital for maintaining balance in marine ecosystems, just as reflexes protect the human body from harm.

CHAPTER 7

CHEMICAL SIGNALING IN THE OCEAN: PREY DETECTION AND THE HUMAN SENSES

The ocean, like a complex living organism, communicates through a vast array of chemical signals. These signals, akin to the sensory systems in the human body, enable marine organisms to detect prey, avoid predators, and navigate their environment. In the same way that our senses—sight, smell, touch, taste, and hearing—help us interpret and respond to the world around us, marine organisms rely on chemical cues to survive in the ocean's vast and often dark waters.

Prey Detection in Marine Ecosystems: The Ocean's Sense of Smell

One of the most fascinating aspects of life in the ocean is the way marine organisms detect their prey. Much like how primitive humans use their sense of smell to detect food or danger, many marine species rely on chemical signaling to locate food sources. Among these chemical signals, dimethylsulfide (DMS) plays a crucial role in the interaction between different levels of the marine food web. DMS is responsible for the characteristic "sea scent" so familiar in coastal areas.

DMS is a sulfur-containing compound produced when phytoplankton are consumed by zooplankton, and its release into the water acts as a chemical cue for a wide variety of marine organisms. Birds such as petrels and shearwaters, for example, use their highly developed sense of smell to detect DMS concentrations in the air, guiding them to areas rich in prey like krill and fish. This is comparable to many terrestrial animals rely on smell to detect the presence of food or certain environmental conditions, though the marine organisms' "sense of smell" is mediated through water and airborne chemical cues rather than direct olfaction as in mammals.

In the same way that human olfactory receptors detect volatile compounds to trigger a response (e.g., if you are not vegan, the smell of a burger may start the salivation process in your mouth), marine predators use specialized receptors to "sniff" out DMS and other compounds in the water. This allows them to track their prey even over long distances, as the chemicals act as beacons that reveal feeding hotspots in the vastness of the ocean.

DMS is not only a marker for prey but also plays a broader role in marine ecological interactions, much like how pheromones in humans and other animals influence behavior. In humans, pheromones are often linked to subtle chemical signals that influence social interactions, such as attraction or territorial behavior. Similarly, DMS can be seen as the ocean's version of a pheromone, guiding marine organisms toward potential food sources and shaping interactions within the food web. The presence of DMS reflects not only the health of marine ecosystems but also how tightly interconnected chemical signaling is with ecological dynamics.

Chemoreception: The Ocean's Sense of Taste and Touch

In the ocean, chemoreception—the ability to detect chemical changes in the environment—is a critical sensory tool for many species. This ability functions much like a combination of the human senses of taste and touch, where organisms detect the presence of chemicals in their surroundings, either through direct contact or by sampling the water. Some fish, for instance, have taste receptors located not only in their mouths but also on their skin and fins, allowing them to "taste" the water for signals of food, mates, or danger. This is analogous to how humans use touch receptors in their skin to sense heat, pressure, or changes in texture, and taste receptors in the mouth to identify food.

Some species, like sharks, take chemical sensitivity to another level with an acute sense of teste. Sharks can detect a single drop of blood in vast quantities of water, thanks to specialized chemoreceptors that allow them to sense minute chemical changes. This is much like the human ability to detect faint flavors, though the precision and distance over which sharks can sense these changes is far more advanced. Additionally, lobsters and octopuses use chemoreception to detect mates and food sources through chemical signals in the water. In coral reefs, chemical cues help guide larvae of corals and fish back to suitable settlement sites after their pelagic stage. These chemical signals, like a "compass," ensure proper reproduction and species distribution across the ecosystem. Chemosensory detection also plays a crucial role at the plankton level, where microorganisms like phytoplankton and zooplankton use chemical cues to sense their environment. Zooplankton utilize chemoreception to locate food sources, avoid predators, and find mates. Some species of microzooplankton can detect and respond to chemical signals released by prey or toxic algae, much like the way larger marine species use chemosensory information to survive and thrive in their environments. These mechanisms illustrate how marine animals rely on the "taste" of their surroundings to make critical life decisions, maintaining ecological balance.

For marine animals, the chemical environment is rich with information. Chemicals released by stressed or injured prey can signal nearby predators, similar to how certain scents can trigger an emotional or behavioral response in humans. For example, the scent of food cooking might make us hungry, while the smell of smoke can trigger a sense of danger. In the ocean, chemical signals have a similar effect, altering the behavior of predators and prey in response to environmental cues.

Drawing of a shark

Vision *vs.* Chemosensing: Detecting Prey in Darkness

While humans primarily rely on vision as the dominant sense for detecting and interacting with the world, many marine organisms must rely more heavily on chemosensing due to the often-dark environment of the ocean's depths. In the epipelagic zone, where sunlight penetrates, vision still plays a significant role, but below this layer, organisms must depend on their ability to detect chemical signals in the water.

Bioluminescent organisms rely primarily on their light-producing abilities to attract mates, signal to each other, or lure prey, rather than using chemical cues in these cases. However, bioluminescence can work in tandem with other sensory mechanisms like chemoreception in some organisms, such as when chemical signals help them detect nearby prey or mates in the dark. The analogy to humans using smell and taste when visibility is poor works well to highlight the sensory adaptability of marine organisms.

In deeper or more turbid waters, the chemical environment becomes the primary mode of communication, much like how the sense of touch becomes more important for humans in situations where vision is impaired. For marine creatures, the ocean's chemical signals are always present, guiding their behaviors and interactions in the same way humans rely on sensory input to navigate their world.

Environmental Stress: Sensory Overload

Just as human senses can be overwhelmed or dulled by environmental factors—such as smoke, noise, or pollution—the chemical signaling in the ocean is also susceptible to disruption by changes in the marine environment. Ocean acidification, for instance, affects the ability of marine organisms to detect and respond to chemical cues, much like how environmental pollutants can interfere with human senses of smell or taste.

Research has shown that elevated carbon dioxide levels can disrupt the sensory systems of fish, impairing their ability to detect predators and navigate their environment. This sensory confusion, brought on by chemical changes in the water, can lead to a breakdown in the natural processes that govern predator-prey interactions, much like how human senses can be overwhelmed by strong odors or loud noises, impairing judgment and reaction times.

Noise and light pollution have increasingly harmful effects on marine animals, disrupting their natural behaviors and ecosystems. Noise pollution, primarily caused by shipping, underwater construction, and military activities, interferes with the communication, navigation, and mating rituals of marine species, particularly affecting animals like whales, dolphins, and fish that rely on sound for echolocation and communication over vast distances. This heightened noise can lead to disorientation, stress, and even stranding in marine mammals.

Luminic pollution, or excessive artificial light from coastal cities and offshore platforms, disrupts the natural light cycles that many marine animals depend on for navigation, reproduction, and feeding. For example, sea turtles, which use the moonlight to navigate to shorelines, are often disoriented by bright city lights, leading to decreased survival rates of hatchlings. Also, zooplankton can alter their daily migratory patterns because of light pollution. These forms of pollution fragment habitats, reduce biodiversity, and alter critical behaviors, posing a growing threat to the health and balance of marine ecosystems.

CHAPTER 8

THE SKELETON OF THE SEA: SILICEOUS AND CALCAREOUS ORGANISMS AS BONE BUILDERS

Just as the human skeleton provides structure, protection, and support for the body, the ocean has its own form of "skeleton" that shapes its ecosystems and habitats. This skeletal framework is formed by marine organisms that build complex structures—ranging from coral reefs to the tiny siliceous and calcareous shells of plankton. These organisms not only create habitats for marine life but also play a crucial role in the geological and chemical processes of the ocean.

In this chapter, we will explore the organisms that form the ocean's skeletal systems, focusing on how their structures provide physical support to marine ecosystems. From the towering coral reefs that act as nurseries for countless species to the microscopic diatoms and coccolithophores that build protective shells, these organisms serve as the structural foundation of the sea. Just as bones provide the human body with shape, stability, and protection, these marine "builders" form the backbone of the ocean's biological and physical landscape.

Coral Reefs: The Ocean's Living Bones

Coral reefs stand as one of the most remarkable and essential skeletal systems in the marine world, forming vast underwater habitats that support an extraordinary array of life. Approximately one-quarter of all marine species, at some point in their life cycle, depend on coral reefs for shelter, food, or breeding grounds. The formation of these reefs begins with coral polyps, small, soft-bodied organisms that secrete calcium carbonate, a substance that hardens to form a protective exoskeleton. Over time, as each generation of polyps builds upon the remains of the previous one, immense coral reefs take shape, creating complex ecosystems that provide essential services to a vast number of marine species. These reefs, akin to the human skeleton, act as the physical backbone of their environment, offering protection and structure in much the same way bones support and protect the human body.

Coral reefs do more than serve as physical habitats. They actively influence their surrounding environment by regulating water flow, stabilizing sediment, and promoting nutrient cycling. The towering coral structures slow ocean currents, creating calm, sheltered waters where a diversity of marine species can thrive. This ability to shape and manage their environment mirrors the role of bone cells in the human body, where the deposition of calcium forms the skeletal structure that supports and protects vital organs. Coral reefs are essential to maintaining ecological balance, providing a foundation for entire ecosystems. Beyond their biological role, they also protect coastlines from the impact of waves and storms. Acting as natural barriers, coral reefs dissipate wave energy, reduce coastal erosion, and mitigate the effects of rising sea levels, much like the human ribcage shields vital organs from external damage.

Diatoms: The Siliceous Skeletons of the Sea

While coral reefs dominate shallow, sunlit waters, another form of skeletal structure thrives within the microscopic realm of plankton. Diatoms, a group of phytoplankton, construct intricately designed silica-based cell walls that not only protect them from predation but also contribute significantly to the marine food web and global nutrient cycling. These single-celled organisms are capable of producing highly efficient glass-like structures known as frustules, which allow them to thrive in a wide range of oceanic environments, from the surface to the depths of the sea. Diatoms play an essential role in photosynthesis, converting sunlight and carbon dioxide into organic matter, forming the foundation upon which much of marine life depends.

The silica shells of diatoms are analogous to the skeletal structures of vertebrates, providing both defense and support. Just as the human skeleton offers structure and protects vital organs, the frustules of diatoms act as both a protective shield and a mechanism that enables them to perform essential functions like photosynthesis. Furthermore, diatoms contribute significantly to the ocean's ability to regulate carbon. When these organisms die, their silica-based shells sink to the ocean floor, carrying with them the carbon absorbed during photosynthesis, by this way effectively reducing atmospheric CO_2 levels and regulating the Earth's climate.

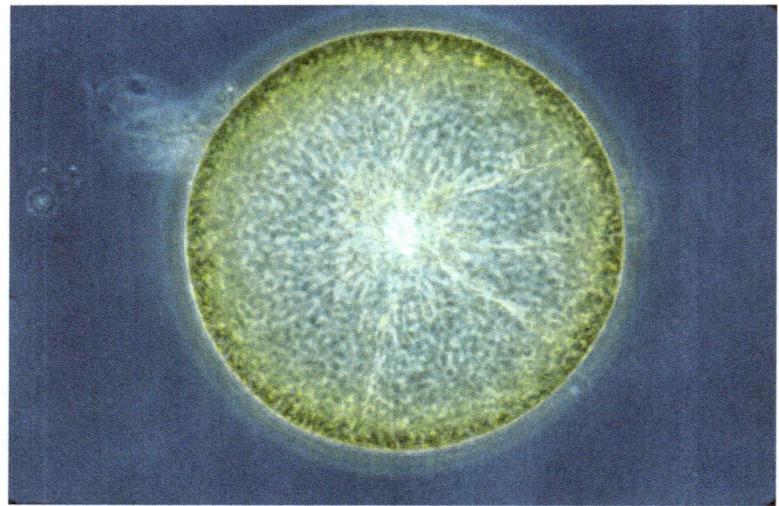

Marine diatom

Coccolithophores: The Calcium Carbonate Architects

Similar to diatoms, coccolithophores, another group of microscopic plankton, also play a crucial role in marine ecosystems by producing calcium carbonate plates called coccoliths. These tiny, disk-shaped plates come together to form spherical exoskeletons that shield coccolithophores from environmental stressors and predation. This protective "armor" requires substantial amounts of energy and calcium, making coccolithophores significant contributors to the ocean's calcium cycle. Their ability to construct these intricate calcium carbonate shells is reminiscent of the mineralization process in vertebrates, where bones provide both structure and protection.

When coccolithophores die, their calcium carbonate exoskeletons sink to the seafloor, where they accumulate as marine sediment. Over time, these sediments form geological structures such as limestone and chalk, with famous formations like the White Cliffs of Dover composed of the remains of coccolithophores that lived millions of years ago. This process is akin to how bones provide long-term support in the human body, ensuring stability and protection over time. The coccolithophores' contribution to marine sedimentation demonstrates their role not only in marine ecosystems but also in shaping Earth's geological landscape. Their calcium carbonate deposits create a lasting legacy, much like how bones contribute to the human body's enduring structure.

Marine Sediments: The Ocean's Geological Framework

The remains of siliceous diatoms and calcareous coccolithophores form a significant part of the ocean's "geological skeleton," accumulating on the seafloor as marine sediments. These layers of sediment act as both habitats for benthic organisms and crucial components of Earth's biogeochemical cycles. Marine sediments provide a rich record of Earth's climate history, preserving evidence of past environmental conditions. The sedimentation process itself plays an essential role in nutrient cycling, as the remains of dead plankton sequester carbon and silica, helping to regulate ocean chemistry and the global carbon balance.

Much like the human skeleton serves as a reservoir of minerals, marine sediments store vital elements like carbon, silica, and calcium, contributing to the overall health of marine ecosystems. These sediments also provide a stable habitat for a wide range of benthic organisms, from worms to crustaceans, which play a critical role in breaking down organic matter and recycling nutrients. The relationship between marine sediments and benthic ecosystems parallels the interaction between bones and the muscles, tendons, and ligaments they support. Just as the skeleton provides a stable framework for movement and function in the human body, marine sediments offer a foundation for the seafloor's biological processes, supporting the productivity and health of marine ecosystems.

The Water Column: The Ocean's Structural Support System

In the human body, the skeleton provides the physical framework that supports muscles, organs, and tissues, enabling movement, protection, and the maintenance of bodily form. In the open ocean, the water column serves a similar structural role, though in a very different way. Rather than a rigid framework, the ocean's water column—defined by layers of varying temperature, density, and pressure—creates a dynamic, fluid structure that supports marine life across different depths. This vertical structure of the ocean, much like the skeleton in a human body, determines the distribution, behavior, and physiological functions of organisms, shaping their ability to thrive in a constantly shifting environment.

The water column's physical properties vary significantly with depth. At the surface, water is warmer, less dense, and subject to the mixing effects of wind and waves. As one descends deeper into the ocean, water becomes colder, denser, and more pressurized. In many ways, this stratified structure is analogous to the skeleton's support system for various organs and tissues, each of which relies on the framework provided by bones. Just as bones allow for the attachment of muscles and the protection of organs, the water column's layers offer different habitats, from the sunlit surface where photosynthesis occurs to the dark, high-pressure environments of the deep ocean where specialized life forms have adapted to extreme conditions.

One of the most significant effects of the water column is buoyancy. Marine organisms do not need rigid skeletal structures like land animals because the density of water provides natural support. Fish, for instance, have evolved swim bladders that help them regulate buoyancy, allowing them to remain suspended at various depths without expending much energy. This is similar to how bones allow terrestrial animals to stand upright, supporting the body's weight against gravity. In the ocean, the water itself acts as a medium of support, reducing the need for heavy skeletons. This buoyancy frees marine organisms from the constraints of gravity, allowing for a wider range of shapes and structures, from the delicate, gelatinous bodies of jellyfish to the rigid, protective shells of mollusks.

Water pressure is another critical factor in the structural dynamics of the ocean's water column. For every 10 meters of depth, pressure increases by about one atmosphere, creating a vastly different environment in the deep ocean compared to surface waters. Organisms living in these extreme depths, such as those near hydrothermal vents or in abyssal plains, have evolved special adaptations to withstand this intense pressure. Some deep-sea fish, for instance, have flexible bodies without gas-filled swim bladders, preventing them from being crushed by the immense weight of the water above. This is analogous to the way bones in the human body are designed to bear weight and protect vital organs from external forces, such as impacts or pressure. The water column, in this sense, provides not only support but also an environment where organisms must develop specific adaptations to survive in the face of immense pressure.

Density and temperature gradients in the water column also play a key role in shaping marine ecosystems. In the human body, the skeleton divides and supports different systems—such as the rib cage protecting the lungs or the spine supporting the central nervous system—allowing for specialized functions in various parts of the body. Similarly, the ocean is divided into distinct zones: the epipelagic (sunlit) zone, the mesopelagic (twilight) zone, the bathypelagic (midnight) zone, and the abyssal zone. Each of these layers provides unique environmental conditions, such as varying levels of light, temperature, and nutrient availability, which in turn determine the distribution of life. Phytoplankton, for instance, thrive in the sunlit upper layers, where they can perform photosynthesis, while bioluminescent creatures and specialized predators like giant squid inhabit the deeper, darker waters.

The stratification of the ocean also governs important processes like nutrient cycling and energy flow. Similar to how bones store and release essential minerals such as calcium and phosphorus, the water column plays a role in cycling nutrients between layers. When surface waters are enriched by sunlight, phytoplankton grow and multiply, forming the base of the marine food web. As they die and sink, their organic material is carried into deeper waters, where it is broken down by decomposers. The movement of nutrients through the water column ensures that marine ecosystems remain productive and balanced, just as the skeleton helps maintain mineral balance within the human body.

However, just as the human skeleton can experience damage or wear over time, the ocean's water column is increasingly affected by human activities, particularly climate change. Rising global temperatures are causing ocean stratification to increase, as warmer, lighter water remains at the surface while deeper, cooler layers are cut off from mixing. This disrupts the natural exchange of nutrients, leading to less productive ecosystems and altering the distribution of marine life. Furthermore, the acidification of the ocean, driven by the absorption of excess carbon dioxide, is affecting the density and chemistry of seawater, making it harder for organisms like corals and shellfish to build their calcium carbonate skeletons. Just as a weakened skeleton in humans can lead to fragile bones and compromised function, these changes in the water column threaten the structural integrity of marine ecosystems.

CHAPTER 9

THE LUNGS OF THE OCEAN: PLANKTON AND OXYGEN PRODUCTION

The lungs are responsible for gas exchange, taking in oxygen and expelling carbon dioxide, a process essential for sustaining life. Similarly, the ocean, often referred to as the "lungs of the planet," plays a critical role in regulating the Earth's atmosphere by producing oxygen and absorbing carbon dioxide (CO_2). At the heart of this process are marine organisms, particularly phytoplankton. These tiny organisms produce nearly half of the oxygen on Earth and play a vital role in carbon sequestration, making them indispensable to both marine and terrestrial life.

In this chapter, we will explore how marine plankton, along with other marine organisms, function as the "lungs" of the ocean, facilitating the exchange of gases that sustains life on Earth. We will examine the process of photosynthesis in phytoplankton, the role of marine respiration, and how the ocean acts as a carbon sink, helping to regulate the global climate. Finally, we will consider the challenges that human activities, such as climate change and ocean acidification, pose to the ocean's ability to continue performing this essential function.

Phytoplankton: The Ocean's Primary Oxygen Producers

Phytoplankton, often referred to as the "grass of the sea," play a crucial role in both marine ecosystems and the global oxygen cycle. These single-celled organisms perform photosynthesis in the sunlit layers of the ocean, converting carbon dioxide and water into glucose and oxygen. However, most of the oxygen produced by phytoplankton is consumed within the ocean by marine organisms, including fish and invertebrates, which rely on dissolved oxygen for respiration. Only in certain productive areas a small portion of the oxygen escapes into the atmosphere, contributing to the air we breathe.

While their oxygen production is essential, phytoplankton are also the foundation of the ocean's food web, supporting marine life from the smallest plankton to the largest predators. Without phytoplankton, marine ecosystems would collapse, as they provide both energy and oxygen to sustain oceanic life. Phytoplankton's photosynthetic process, similar to that of terrestrial plants, is vital to life on Earth, maintaining the balance of oxygen in marine environments and ensuring the health of the broader biosphere.

Marine Respiration: The Ocean's Carbon Dioxide Exhalation

While phytoplankton produce oxygen and absorb carbon dioxide through photosynthesis, marine organisms also "exhale" carbon dioxide through the process of respiration. Much like how humans inhale oxygen and exhale carbon dioxide as a byproduct of cellular respiration, marine organisms—including fish, zooplankton, and even phytoplankton themselves—consume oxygen and release CO_2 as they break down organic molecules for energy. This constant exchange between oxygen production and carbon dioxide release is an essential part of the ocean's metabolic balance, contributing to the broader carbon and oxygen cycles.

In the ocean, respiration is carried out by all aerobic organisms, from microscopic plankton to massive marine mammals, such as whales. As they respire, these organisms use oxygen to metabolize glucose, releasing carbon dioxide and water as byproducts. The carbon dioxide produced during respiration is then released into the surrounding water, where it becomes part of the ocean's carbon cycle. Some of this CO_2 is reabsorbed by phytoplankton for photosynthesis, while the rest may be exchanged with the atmosphere. This dynamic balance between photosynthesis and respiration ensures the continued health of marine ecosystems and maintains the global balance of oxygen and carbon dioxide.

Drawing of whales

The Ocean as a Carbon Sink: Mitigating Climate Change

The ocean plays a crucial role in mitigating the effects of climate change by acting as a carbon sink, absorbing and storing vast amounts of carbon dioxide from the atmosphere. In addition to the biological pump, which transfers organic carbon to the deep ocean, the ocean absorbs CO_2 through a process known as dissolution, where carbon dioxide dissolves into seawater and becomes part of the inorganic carbon cycle. This dissolved carbon can be used by phytoplankton for photosynthesis or remain in the water, contributing to the long-term storage of carbon in the ocean.

The ocean currently absorbs approximately 25% of human-induced CO_2 emissions, helping slow global warming. However, as atmospheric CO_2 levels continue to rise (now exceeding 420 ppm), the ocean's ability to absorb carbon is diminishing. Increased ocean acidification has already reduced the average pH of the ocean from 8.2 to around 8.1, which, though seemingly small, represents a 30% increase in acidity. Even slight shifts in pH can disrupt calcium carbonate formation for species like corals, mollusks, and plankton, weakening their ability to build shells and skeletons, threatening marine ecosystems.

In addition to acidification, rising sea temperatures are reducing the solubility of CO_2. For every 1°C rise in ocean temperature, the ability of seawater to absorb CO_2 decreases, contributing further to the buildup of greenhouse gases in the atmosphere. This warming is also increasing ocean stratification, preventing the mixing of nutrient-rich deep waters with surface layers. This inhibits phytoplankton growth, which is critical for carbon sequestration through the biological pump—a process where organic carbon sinks to the ocean depths.

Without the crucial role of phytoplankton, which currently absorb an estimated 10 gigatons of CO_2 annually, the ocean's carbon sink capacity diminishes, destabilizing the marine food web. Human activities, such as deforestation, burning of fossil fuels, and industrial agriculture, continue to drive the rise in CO_2, with global carbon emissions in 2021 reaching 36.3 billion metric tons. If the ocean's capacity to absorb CO_2 continues to decline, it will accelerate the pace of climate change, exacerbating rising temperatures, sea-level rise, and extreme weather events.

CHAPTER 10

THE REPRODUCTIVE SYSTEM: MARINE LIFE CYCLES AND REPRODUCTION

Our reproductive system is vital for ensuring the continuation of the species, passing on genetic material to future generations. The ocean, too, relies on the reproductive success of its diverse inhabitants to maintain biodiversity, replenish populations, and sustain the intricate web of life that makes up marine ecosystems. From the simple, asexual reproduction of plankton to the complex life cycles of fish and marine invertebrates, the reproductive strategies of ocean species are as varied as they are fascinating.

In this chapter, we will explore the reproductive systems of marine organisms, focusing on the unique and complex ways in which plankton, fish, and other marine species ensure the survival of their populations. We will examine the diverse strategies of reproduction, from asexual reproduction in phytoplankton to the elaborate mating rituals of marine animals. Additionally, we will explore how reproductive processes in the ocean are influenced by environmental conditions and human activities, and how the disruption of these systems can have far-reaching impacts on marine biodiversity.

Asexual Reproduction: Phytoplankton and Rapid Population Growth

Phytoplankton, the primary producers of the ocean, rely primarily on asexual reproduction to sustain their populations. This mode of reproduction allows phytoplankton to respond quickly to favorable environmental conditions, such as nutrient upwelling or seasonal changes in sunlight. Asexual reproduction in these microscopic organisms is usually achieved through binary fission, a process in which a single phytoplankton cell divides into two genetically identical daughter cells. This ability to reproduce rapidly through binary fission enables phytoplankton populations to multiply quickly when conditions are ideal. In some cases, these divisions can occur in as little as a few hours, leading to the formation of massive blooms that can stretch across vast areas of the ocean's surface.

The rapid reproductive capacity of phytoplankton is similar to the regenerative abilities in the human body, where cells divide and multiply to repair tissue or sustain growth. Just as humans rely on cell division to heal from injury or to grow, phytoplankton exploit favorable conditions to regenerate and expand their populations. This allows them to play their critical role as the foundation of the marine food web, supporting zooplankton, fish, and larger marine animals that depend on them for sustenance. However, while the ability to reproduce quickly ensures that phytoplankton can take advantage of optimal conditions, their lack of genetic diversity due to asexual reproduction makes them vulnerable to environmental changes, such as temperature fluctuations or pollution, in much the same way that limited genetic diversity in human populations can increase susceptibility to disease.

Sexual Reproduction in Zooplankton and Marine Invertebrates

While phytoplankton primarily reproduce asexually, metazoan zooplankton and many marine invertebrates, such as mollusks and crustaceans, engage in sexual reproduction, a process that introduces genetic diversity and enhances the ability of species to adapt to environmental changes. The fusion of male and female gametes during sexual reproduction results in offspring that carry genetic traits from both parents, fostering genetic variability that can help species survive in fluctuating conditions. Many marine organisms exhibit complex

reproductive strategies that are intricately linked to environmental cues such as changes in temperature, lunar cycles, and food availability.

One fascinating reproductive strategy in marine life is broadcast spawning, where marine invertebrates, including corals, sea urchins, and clams, release large quantities of eggs and sperm into the water column simultaneously. This mass release saturates the water with gametes, increasing the likelihood of fertilization. Such events are often synchronized with external cues like the full moon or seasonal temperature shifts, ensuring that eggs and sperm are present in the water at the same time. Coral reefs, in particular, showcase the spectacle of mass spawning, where millions of eggs and sperm fill the water in a flurry of reproductive activity, often creating a "snowstorm" effect. This synchrony is crucial for the survival of coral species, as it increases the chances of fertilization and subsequent larval development. The reliance on such environmental cues for reproductive success is not dissimilar to how hormonal cycles in humans regulate reproductive processes and the timing of key physiological events.

Zooplankton, including species such as copepods, krill, and jellyfish, have intricate life cycles that involve distinct larval stages. Many undergo metamorphosis, transitioning from one developmental stage to another as they grow and adapt to new ecological roles. Most jellyfish, for example, begin their lives as tiny planula larvae before settling on the ocean floor and transforming into polyps. Eventually, these polyps metamorphose into the familiar medusa form, capable of free-swimming and reproduction. This transformation echoes the various stages of human growth and development, where individuals undergo significant changes from infancy to adulthood. Just as humans transition through different life stages to fulfill different societal roles, zooplankton evolve through multiple stages to take on their ecological functions in the marine environment.

Jellyfish

Marine Fish Reproduction: Complex Strategies and Larval Dispersal

Marine fish exhibit a wide array of reproductive strategies, from external fertilization through spawning to more elaborate forms of parental care. The reproductive success of fish is intimately tied to the survival of their larvae, many of which rely on ocean currents for dispersal, ensuring that they can reach suitable habitats and food sources. Environmental factors such as temperature, salinity, and habitat availability play a significant role in shaping fish reproductive strategies, dictating when and where fish choose to spawn.

External fertilization, where females release eggs into the water and males release sperm to fertilize them, is common among many fish species. This method, often timed with environmental signals, allows for the rapid dispersal of larvae into the surrounding waters. Salmon, for example, embark on long migrations from the ocean to freshwater rivers to lay their eggs in gravel beds, ensuring that the eggs are protected from predators and strong currents. Once hatched, the larvae are often carried by ocean currents to new habitats, where they can find food and shelter. This larval dispersal is essential for maintaining genetic diversity and supporting the colonization of new areas. In much the same way that human offspring depend on care and protection in their early years, marine fish larvae rely on the availability of suitable nursery habitats and abundant food to ensure their survival.

Some fish species, such as seahorses and clownfish, exhibit more complex forms of parental care, providing a degree of protection to their offspring after fertilization. Seahorses, for instance, are known for the unique role males play in reproduction; the fertilized eggs are carried in a specialized brood pouch until they hatch, offering a safe and nourishing environment for the developing embryos. Similarly, cichlids and clownfish guard their eggs, protecting them from predators and environmental stressors. This level of parental care can be compared to the care human parents provide for their young, ensuring the survival and well-being of the next generation.

Reproduction Challenges and Adaptations of Marine Mammals

Marine mammal reproduction mirrors some aspects of human birth and infancy, but with unique adaptations to the ocean's environment. Just as human newborns are entirely dependent on their parents for survival, marine mammal calves are highly reliant on their mothers from the moment they are born. However, unlike humans, marine mammals must be born directly into the water—a vast and often dangerous environment. To ensure their survival, many marine mammals give birth tail-first, a precaution that prevents the newborn from drowning, much like how human babies immediately begin breathing air after birth.

Nursing is another area of both similarity and difference. Human babies feed frequently to gain weight and grow, but marine mammal mothers face the added challenge of feeding their young while swimming and avoiding predators. To compensate, marine mammal milk is extremely rich in fat—up to 10 times more so than human milk—allowing the calf to gain weight rapidly and build the strength needed to navigate the ocean. This accelerated growth is crucial, much like a human infant's early developmental milestones, but in the ocean, these milestones include learning to swim and dive at an early age.

The bond between mother and calf is also essential, as in humans. Communication, whether through vocalizations in whales and dolphins or physical contact in seals, helps mothers protect their young from dangers in the deep. Similar to how human parents must teach their children essential skills, marine mammal mothers guide their calves through critical behaviors such as finding food, avoiding predators, and navigating long migrations. Yet, unlike humans, marine mammal calves must learn these survival skills in a far more demanding environment, facing the constant threat of predators and the challenges of life in a vast, ever-changing ocean.

CHAPTER 11

THE EXCRETORY SYSTEM: PLANKTON AND NUTRIENT RECYCLING

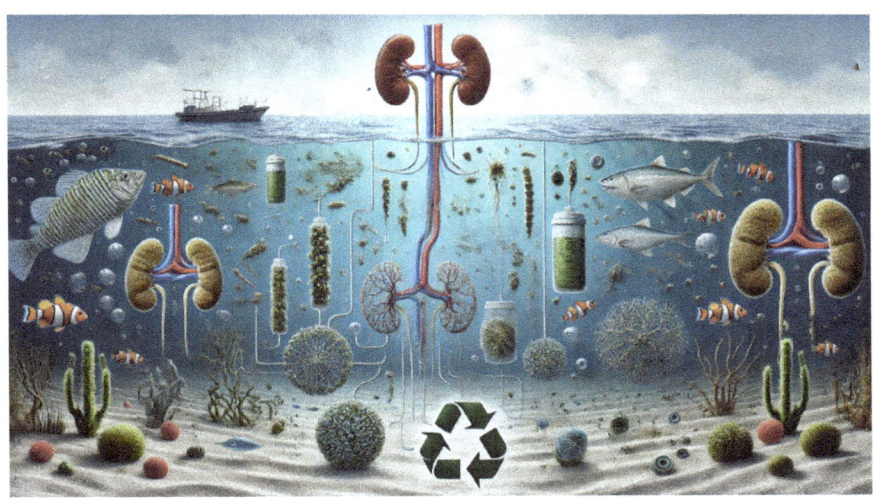

The excretory system plays a vital role in maintaining the balance of internal fluids by filtering out waste products and recycling essential nutrients. The ocean, as a vast and interconnected system, has its own version of an excretory process, whereby marine organisms—including plankton, bacteria, and other creatures—recycle nutrients and process waste to maintain the health of marine ecosystems. This nutrient recycling is essential for sustaining life in the ocean, ensuring that vital compounds such as nitrogen, phosphorus, and carbon remain available for primary production and growth.

In this chapter, we will explore how marine organisms contribute to the ocean's excretory system, focusing on the role of plankton and microbial life in nutrient recycling and waste processing. We will examine the microbial loop, the biogeochemical cycles of nitrogen and phosphorus, and how these processes maintain the balance of marine ecosystems. Additionally, we will discuss the impact of human activities—such as pollution and overfishing—on the ocean's ability to recycle nutrients, as well as the growing problem of eutrophication and dead zones caused by nutrient imbalances.

The Role of Plankton in Nutrient Recycling

Plankton serve as the primary drivers of nutrient cycling in marine ecosystems. Phytoplankton capture sunlight and convert carbon dioxide into organic matter through photosynthesis, providing the foundation for much of the marine food web. Zooplankton, as primary consumers, feed on phytoplankton and release waste products back into the surrounding waters. These waste materials, alongside the organic matter from dead organisms, are further broken down by bacteria and other microorganisms, ensuring that nutrients are continuously recycled. This constant recycling makes nutrients available to phytoplankton and other marine organisms, maintaining the productivity and health of the ocean.

A key component of this nutrient cycling process is the microbial loop, which involves bacteria and other microorganisms breaking down dissolved organic matter and converting it back into inorganic nutrients, such as nitrogen and phosphorus. This cycle is crucial for sustaining the productivity of marine ecosystems. Without it, essential nutrients would be depleted quickly, limiting the growth of phytoplankton and, consequently, the entire food web. In the microbial loop, organic matter, including dead organisms and fecal pellets, is consumed by heterotrophic bacteria that decompose the material into simpler compounds. These bacteria are, in turn, consumed by small protozoans and other microorganisms, creating a continuous recycling loop. The microbial loop can be compared to the human body's kidneys and liver, which filter and process waste while retaining essential nutrients. In both systems, this recycling ensures that nutrients are not wasted and that the system remains balanced and functional.

Zooplankton also play a direct role in the release of nutrients back into the water. As they feed on phytoplankton and other organic matter, zooplankton excrete ammonium, phosphate, and other dissolved inorganic nutrients, which are then taken up by phytoplankton. This process, known as regeneration, helps to sustain primary production in nutrient-poor regions of the ocean. Zooplankton act as conduits for the transformation of nutrients from organic matter back into the ecosystem. In a way, they function similarly to the human excretory system, which recycles water, electrolytes, and nutrients while eliminating waste. Without this continual recycling, essential nutrients in the ocean would quickly become scarce, limiting the growth of phytoplankton and disrupting the marine food chain.

Copepods are the most abundant group of zooplankton

The Nitrogen Cycle: A Key Component of Marine Productivity

The nitrogen cycle is one of the most critical biogeochemical cycles in the ocean because nitrogen is essential for the growth of marine organisms. Nitrogen is a fundamental component of amino acids, proteins, and nucleic acids, all of which are necessary for life. Despite its abundance in the atmosphere as nitrogen gas (N_2), most marine organisms cannot use nitrogen in this form. Instead, they rely on dissolved inorganic nitrogen compounds, such as ammonium, nitrate, and nitrite, which are produced through a series of biological processes.

The process of nitrogen fixation plays a central role in making nitrogen available to marine ecosystems. Certain bacteria and cyanobacteria, which are a type of phytoplankton, can convert atmospheric nitrogen into ammonium, a form that can be used by other organisms. These nitrogen-fixing organisms are especially important in regions of the ocean where other forms of nitrogen are limited. Nitrogen fixation can be likened to the human body's need for certain nutrients, such as vitamins and minerals, to support essential biological functions. Without nitrogen-fixing bacteria, the availability of nitrogen in the ocean would be severely restricted, reducing the productivity of phytoplankton and, by extension, the entire marine food web.

Once ammonium is present in the water, it can be converted into nitrate through nitrification, a process carried out by specialized bacteria. Nitrate is the most readily available form of nitrogen for diatoms and some other microalgae, which use it for growth and reproduction. After nitrogen has cycled through the marine food web, it can be returned to the atmosphere through denitrification, where bacteria convert nitrate back into nitrogen gas. This continuous cycling of nitrogen is similar to how the human body breaks down proteins, recycles amino acids, and removes excess nitrogen through the urea. Both processes ensure that vital nutrients are not lost and that biological systems remain productive and efficient.

The Phosphorus Cycle: Supporting Marine Life

Phosphorus is another essential nutrient in marine ecosystems, required for the formation of DNA, RNA, and ATP[6]—the molecule that provides energy for cellular processes. Unlike nitrogen,

---

[6] DNA: Deoxyribonucleic Acid, RNA: Ribonucleic Acid, ATP: Adenosine Triphosphate

phosphorus does not have a gaseous phase, and its availability in the ocean is largely dependent on the input of phosphate from rivers, the erosion of rocks, and the decomposition of organic matter. Phosphate is readily taken up by phytoplankton and other primary producers, who incorporate it into their biomass. When these organisms die or are consumed, bacteria and other microorganisms break down their organic matter, releasing phosphate back into the water.

The recycling of phosphorus is crucial for maintaining the productivity of marine ecosystems, as it supports the growth of phytoplankton and other organisms. This process can be compared to how the human body regulates phosphate levels through the kidneys, ensuring that cells have the energy needed to perform their essential functions. Without the continuous cycling of phosphorus, marine ecosystems would struggle to sustain life, particularly in regions where nutrient inputs are limited.

A significant portion of phosphorus is stored in marine sediments, where it can remain for extended periods before being re-released into the water column through upwelling or other processes. These sediments act as nutrient reservoirs, helping to regulate the availability of phosphorus in the ocean. The role of marine sediments in phosphorus cycling is akin to the way the human body stores minerals like calcium in bones, which can be mobilized when needed to maintain metabolic balance. In both systems, these nutrient reserves ensure that essential compounds are available for future use, supporting long-term productivity.

Liver, Kidneys, and Estuaries: The Ocean's Detox and Filtration Systems

The liver in the human body plays a critical role in detoxifying blood, processing nutrients, and removing harmful substances. Likewise, kidneys function as essential filters, regulating the body's fluid balance by removing waste products and excess nutrients through urine. Together, these organs maintain the body's internal environment by ensuring that toxins are eliminated and that nutrient levels remain balanced. Similarly, estuaries—the dynamic zones where freshwater from rivers meets the ocean—act as natural filtration and detox systems for marine environments.

Estuaries serve as both the liver and kidneys of the ocean. Like the liver, estuaries detoxify the ocean by trapping pollutants such as heavy metals, agricultural runoff, and sediments. These pollutants are filtered by marsh plants, filter-feeding organisms like oysters and clams, and sedimentation processes within the estuarine environment. This prevents harmful contaminants from reaching the open ocean, where they could disrupt marine ecosystems. Just as the liver neutralizes toxins before they can spread throughout the body, estuaries cleanse the water before it flows into the sea, protecting marine life.

In addition to detoxification, estuaries regulate nutrient levels, much like the kidneys manage fluid and electrolyte balance in the body. Estuaries help process and recycle nutrients such as nitrogen and phosphorus, ensuring they are available in the right amounts for marine organisms to thrive. By controlling the flow of nutrients and reducing excess, estuaries prevent issues like eutrophication, where too many nutrients in the water can lead to harmful algal blooms and oxygen-depleted zones, which can suffocate marine life. This mirrors how the kidneys maintain homeostasis by regulating the balance of salts, water, and waste products, ensuring the body's environment remains stable and healthy.

Estuaries also function as nurseries for many marine species, providing a nutrient-rich and protected environment where young fish and invertebrates can grow before moving into the open ocean. This role is analogous to the liver's function in producing proteins and aiding in cell repair, ensuring the body's tissues function optimally. Estuaries not only support the health of individual species but also sustain the broader marine food web, ensuring the entire ecosystem remains balanced and productive.

In this way, estuaries perform a dual role similar to the liver and kidneys in the human body. They filter harmful substances, regulate nutrient levels, and provide a stable environment for growth, ensuring the health of marine ecosystems much like these vital organs ensure the balance and detoxification of the body. Without healthy estuaries, the ocean would be more vulnerable to pollution, nutrient imbalances, and the collapse of key ecosystems—paralleling the dangers faced by the human body when the liver or kidneys fail to function properly.

Eutrophication and Dead Zones: The Consequences of Nutrient Imbalances

While the recycling of nutrients is vital for sustaining marine ecosystems, excessive nutrient input—particularly nitrogen and phosphorus—can have harmful consequences. Eutrophication occurs when nutrient levels in the water become too high, often as a result of agricultural runoff, sewage discharge, or industrial pollution. The excess nutrients fuel the rapid growth of phytoplankton and algae, leading to blooms that can throw marine ecosystems out of balance.

Harmful algal blooms, driven by nutrient overload, produce toxins that can harm marine life and humans. As these blooms die and decompose, the breakdown of organic matter depletes oxygen in the water, creating dead zones—areas where oxygen levels are too low to support most forms of marine life. These oxygen-deprived regions are a growing concern in coastal areas worldwide, including the Gulf of Mexico and parts of the Baltic Sea. The collapse of oxygen levels leads to the death of fish, shellfish, and other marine organisms, causing devastating effects on biodiversity and local economies.

The ocean's excretory system, responsible for processing and recycling nutrients, is increasingly overwhelmed by human activities. The overloading of marine ecosystems with nutrients, combined with the impacts of climate change—such as warming waters and ocean acidification—is disrupting the natural balance of nutrient cycling. This disruption is similar to how kidney disease or liver dysfunction can impair the body's ability to filter waste and maintain chemical equilibrium. Just as the human body depends on a healthy excretory system to remove toxins and regulate nutrients, the health of the ocean depends on its ability to recycle nutrients without becoming overloaded.

CHAPTER 12

THE HORMONAL CYCLES OF THE OCEAN: SEASONALITY AND BIOLOGICAL RHYTHMS

Akin the human body experiences seasonal cycles, dictated by hormonal changes that influence everything from mood to metabolism, the ocean follows its own rhythmic patterns. These oceanic cycles are deeply connected to the Earth's rotation, the tilt of its axis, and the changing seasons, which create variations in light, temperature, and nutrient availability. The similarities between the ocean's seasonal changes and the hormonal cycles in humans reveal a deeper connection between natural rhythms that govern both ecosystems and our own bodies.

In this chapter, we will explore the parallels between the ocean's seasonal shifts and human hormonal cycles, demonstrating how these periodic fluctuations regulate life, productivity, and reproduction in both realms. We will examine how the ocean's patterns of growth, rest, and renewal align with the hormonal cycles that govern human biology, from the rise and fall of energy levels to the regulation of reproductive processes.

Circadian Cycles in the Ocean and Humans: Rhythms of Life

Both humans and marine organisms are governed by circadian rhythms—the internal processes that follow a roughly 24-hour cycle and regulate essential biological functions such as sleep, wakefulness, feeding, and activity. These rhythms are influenced by environmental cues, particularly the cycle of light and darkness, and they play a critical role in maintaining the balance of life. While circadian rhythms are often associated with human sleep-wake cycles, they are just as vital in the ocean, where countless marine organisms, from the smallest plankton to the largest predators, follow daily patterns of behavior.

In humans, circadian rhythms are regulated by the suprachiasmatic nucleus in the brain, which acts as the body's internal clock. This clock is sensitive to environmental light, which signals to the body when to wake up, when to sleep, and when to be most active. Melatonin, a hormone that promotes sleep, increases as darkness falls, while cortisol, a hormone linked to wakefulness and energy, peaks with daylight. This cycle of hormone regulation dictates the body's energy levels, metabolism, and mood, ensuring that biological functions are optimized for the day's activities.

These rhythms are not limited to sleep but also influence other critical processes such as digestion, cognitive function, and physical performance. Disruptions to this natural cycle—such as from irregular sleep schedules, exposure to artificial light at night, or long-distance travel—can lead to health problems, including fatigue, mood disorders, and metabolic issues.

Just as humans rely on these daily cycles for maintaining health, marine organisms also experience circadian-driven behaviors, particularly in relation to light and darkness. For example, the reef fish that hide among corals during the day emerge to hunt at night, taking advantage of reduced predator activity under the cover of darkness. In many cases, entire ecosystems are governed by these rhythms. In tropical regions, for instance, corals synchronize their spawning events with lunar cycles and seasonal cues, ensuring the greatest chance of successful fertilization. These spawning events often occur during the night, with corals releasing billions of eggs and sperm into the water simultaneously, creating a spectacle of life driven by the rhythms of the sea. Other marine organisms, such as nocturnal squid, also follow daily migration patterns. They ascend to the ocean's surface at night to hunt, taking advantage of reduced predator activity, and return to the depths by day. Like zooplankton, their activity is finely tuned to the rhythm of light and darkness, demonstrating how circadian rhythms govern feeding behaviors across a range of species.

However, perhaps the most dramatic examples of circadian rhythms in the ocean is zooplankton vertical migration, the largest migration of animals on the planet, occurring daily. Zooplankton follow a strict daily pattern of movement in response to the light-dark cycle. At dusk, as the sunlight fades, trillions of zooplankton rise from the deep, dark waters to the ocean's surface to feed on phytoplankton. At dawn, before the sun rises, they retreat back into the depths to avoid predators that hunt by sight. This behavior is driven by the need to balance feeding with safety. During the day, visual predators such as fish and seabirds are more active in the well-lit surface waters. To avoid being eaten, zooplankton stay deep below the surface where it is too dark for predators to spot them. However, when night falls, the surface becomes safer, allowing zooplankton to migrate upward and feed on the abundant phytoplankton that thrive in the sunlight.

This daily cycle of movement is essential not only for the survival of zooplankton but also for the entire marine ecosystem. As they move up and down the water column, zooplankton transport nutrients between the deep and surface waters, facilitating nutrient cycling in the ocean. The vertical migration also plays a significant role in the biological pump, a process that we already know sequesters carbon by moving carbon-rich material from the surface to the ocean's depths.

The movement of zooplankton mirrors the human circadian rhythm of wakefulness and rest. Just as humans rise with the sun to begin their daily activities and rest at night, zooplankton rise at night to feed and retreat by day for safety. Both systems rely on a predictable cycle of light and darkness to regulate behavior and energy use. In humans, the cycle dictates when we sleep, eat, and work, while in the ocean, it dictates when zooplankton feed and hide.

For zooplankton, this vertical displacement is closely tied to feeding rhythms. Interestingly, these feeding patterns are not exclusive to zooplankton and fish; organisms across a wide range of sizes, from the tiniest protozoans to the largest whales, exhibit diel feeding rhythms. Most marine species optimize their feeding behavior by synchronizing it with periods of maximum food availability while minimizing exposure to predators.

The Importance of Light in Regulating Daily Rhythms

Both the human body and the ocean rely heavily on light to regulate their respective circadian cycles. For humans, exposure to natural sunlight helps keep the body's internal clock in sync with the external environment, promoting healthy sleep patterns, metabolism, and mood. Blue light from the sun, in particular, suppresses melatonin, keeping us alert and energized during the day. At night, the absence of light allows melatonin to rise, helping the body prepare for sleep.

In the ocean, light plays an equally critical role. The ocean's photic zone, where sunlight penetrates, supports the vast majority of marine life, particularly phytoplankton, which rely on sunlight to produce energy through photosynthesis. Zooplankton time their migrations to avoid the harsh light of day, rising to feed when the darkness of night offers protection. Light not only dictates feeding and activity cycles but also influences reproduction and migration patterns in many marine species.

The Ocean's Annual Cycle: Growth and Renewal

The seasonal cycle of plankton in the ocean is one of the most dynamic and essential processes in marine ecosystems. Plankton experience fluctuations in growth and activity that are closely tied to the changing seasons. These seasonal cycles are driven by variations in light, temperature, and nutrient availability, which together determine when plankton bloom and when they retreat.

Interestingly, this natural rhythm of plankton growth and decline mirrors the human seasonal cycles—the way our energy levels, moods, and even productivity fluctuate throughout the year. As we move through spring, summer, autumn, and winter, our bodies respond to environmental changes much like plankton respond to shifts in their environment. This section explores the parallels between the seasonal cycle of plankton and the human experience throughout the year.

Marine copepods are strong migrators within zooplankton

Spring marks a period of renewal and growth for both plankton in the ocean and humans on land. As the days grow longer and sunlight becomes more abundant, the ocean's surface layers warm up, and the spring bloom of plankton begins. Phytoplankton rapidly reproduce in response to the increased sunlight and availability of nutrients that accumulated during the winter months. This burst of life fuels the entire marine ecosystem, as zooplankton and in turn, become food for fish, seabirds, and marine mammals.

In humans, spring is often associated with a similar renewal of energy and vitality. After the cold and dark winter months, the increasing daylight and warming temperatures signal to our bodies that it is time to become more active. Melatonin levels, which rise during the longer nights of winter, begin to fall, while serotonin and dopamine—hormones linked to mood and energy—start to increase. This leads to a sense of awakening and motivation, much like the plankton's rapid growth in response to sunlight.

Both plankton and humans experience a surge of productivity during this time. For plankton, this is their most active and fertile period, with phytoplankton blooms driving the entire marine food web. For humans, spring often brings a renewed focus on goals, creativity, and physical activity, as our bodies and minds respond to the lengthening days and warmer temperatures. However, just as this high-energy phase does not last forever in plankton, humans also eventually reach a point where growth and energy begin to stabilize.

As spring transitions into summer, both plankton and humans reach a peak of activity. The ocean remains warm, and there is still plenty of sunlight for phytoplankton to continue growing, but by mid-summer, nutrient levels in the surface waters begin to deplete. The plankton bloom, which once thrived on the nutrient-rich waters of spring, starts to slow down as these essential nutrients are used up.

Zooplankton and other marine organisms may still be active, but the rapid growth phase of spring has passed, and the ecosystem begins to experience a gradual decline in productivity.

Similarly, humans often reach a peak of energy and activity during the early summer weeks. Long days, warm weather, and the abundance of social and outdoor activities create a sense of high energy and productivity. However, much like the plankton, humans can also experience a sense of exhaustion as summer progresses. The long hours of daylight and intense activity and the excessive heat of summer may act lowering blood pressure and giving a sensation of fatigue.

As the seasons change from summer to autumn, both plankton and humans experience a period of transition. In the ocean, the thermocline begins to break down. This allows nutrients from the deep to mix back into the surface layers, creating the potential for minor plankton blooms. However, while there are more nutrients available, the light and temperature are no longer optimal, limiting the growth of phytoplankton. These smaller blooms are not as dramatic as the spring bloom, but they provide a final burst of activity before the winter months.

For humans, autumn often brings a similar period of reflection and transition. The longer days of summer give way to shorter daylight hours, and as temperatures cool, people often slow down and turn their focus inward. This time of the year is marked by a shift in energy—while we may still be active, we often begin to conserve energy and prepare for the colder, darker months ahead. Autumn can be a time of productivity, but it is less frenetic than spring, much like the smaller plankton blooms of the season.

Winter brings a stark contrast to the growth and activity of spring and summer. In the ocean, the winter months are characterized by plenty of nutrients but very little sunlight. The surface waters are cold, and the days are short, creating conditions that are not conducive to plankton growth. Phytoplankton production slows to a near halt, and many zooplankton retreat to deeper waters to conserve energy. While the ocean is rich in nutrients, it is simply too dark and cold for significant biological activity to occur. This period of dormancy allows the ocean to rest and restore, building up nutrients in preparation for the spring bloom.

For humans, winter is often a time of restoration and reflection. Shorter days and colder temperatures lead to a natural inclination toward rest (and eating heavy meals, with its consequences), with many people experiencing lower energy levels and a desire for more sleep. Just as the ocean's productivity slows during the winter, humans often experience a period of reduced activity, focusing on conserving energy rather than growth. Circadian rhythms are closely tied to the seasons, and in winter, many people find themselves more in tune with the natural cycle of rest and renewal.

Tidal Cycles: Oceanic Hormones in Motion

Beyond the annual cycle of the seasons, the ocean also experiences tidal cycles, driven by the gravitational pull of the moon. These tidal shifts, which occur on a regular basis, affect the movement of water, the availability of nutrients, and the behavior of marine species. In humans, the hormonal cycles that regulate everything from reproductive health to mood and energy levels function in much the same way.

For example, the menstrual cycle in women, governed by the ebb and flow of hormones like estrogen and progesterone, parallels the ocean's tidal cycles. Just as the tides rise and fall, creating periods of heightened activity and rest for marine ecosystems, the human body follows a rhythm that balances periods of energy and rest. Ovulation, which occurs mid-cycle, is like a spring tide—a peak of fertility and energy.

As hormone levels rise, the body prepares for reproduction, much like how the ocean's tides stimulate the movement of nutrients and the activity of marine life.

In the ocean, the movement of tides also dictates the behavior of many species. Tidal rhythms govern the feeding, breeding, and migration patterns of countless marine organisms, just as hormonal cycles influence human behaviors, moods, and physical well-being. Fish, crabs, and other intertidal species rely on the rhythmic rise and fall of the water to time their reproductive activities, while humans experience similar peaks and troughs of emotional and physical states as their hormone levels fluctuate.

Both the ocean and the human body rely on these rhythmic cycles to maintain balance and ensure the continuation of life. Just as disruptions in hormonal cycles can lead to health problems in humans—such as mood disorders, infertility, or metabolic issues—disruptions in the ocean's tidal cycles, whether through climate change or human interference, can have devastating effects on marine ecosystems.

Hormonal Regulation in the Ocean

In humans, the release of hormones is tightly regulated by signals from the brain's hypothalamus and pituitary gland, which respond to internal and external cues such as light, stress, and nutrition. The ocean's cycles, too, are regulated by external forces—primarily light, temperature, and nutrient availability. The amount of sunlight penetrating the ocean's surface influences the growth of phytoplankton, just as sunlight triggers the production of hormones like melatonin and serotonin in humans.

As the seasons change, the angle of the sun's rays shifts, altering the productivity of marine ecosystems. In the same way that human hormonal cycles are influenced by the amount of daylight we receive, the ocean's seasonal cycles are driven by changes in light and temperature. These changes regulate the timing of plankton blooms, the migration patterns of fish, and the reproductive cycles of marine species. Nutrient levels, replenished during the winter months, act as the ocean's "hormonal reserves," much like how the body stores energy and essential nutrients to fuel future growth and reproduction.

Disruptions to these regulatory cycles can have significant consequences. In humans, stress, poor diet, or lack of sleep can throw off hormonal balance, leading to mood swings, fatigue, or chronic health problems. In the ocean, climate change and pollution are disrupting the seasonal cycles that regulate the health of marine ecosystems. Warmer waters are causing phytoplankton blooms to occur earlier or later than usual, throwing off the timing of the entire food web. Fish and marine mammals, which rely on these cycles for breeding and feeding, are increasingly out of sync with the rhythms of the ocean.

The Importance of Balance: Hormonal Health for the Ocean and Humans

Both the ocean and the human body rely on a delicate balance to function properly. In humans, hormonal imbalances can lead to conditions like PCOS[7], thyroid disorders, and mood disorders. Similarly, imbalances in the ocean's seasonal cycles, driven by human

---

[7] PCOS (Polycystic Ovary Syndrome) is a common hormonal disorder that affects women of reproductive age.

activity, are leading to ecological problems like coral bleaching, ocean acidification, and species displacement.

Just as hormonal health in humans requires care, attention, and a balance of lifestyle factors, the health of the ocean's cycles depends on our ability to protect and preserve its natural rhythms. Overfishing, pollution, and climate change are disrupting the ocean's hormonal-like cycles, pushing ecosystems out of balance and making it difficult for marine life to thrive.

By understanding the parallels between human hormonal cycles and the ocean's seasonal rhythms, we can gain a deeper appreciation for how interconnected these systems are. Just as we take care of our bodies through healthy living, we must care for the ocean by reducing pollution, protecting biodiversity, and addressing the root causes of climate change. Maintaining this balance is essential for both the health of the planet and our own well-being.

CHAPTER 13

DIAGNOSING OCEAN ILLNESS: A PARALLEL TO HUMAN DISEASES

The ocean, like the human body, can fall ill. While the human body experiences fevers, heart attacks, skin diseases, and immune system breakdowns, the ocean too is facing its own set of crises, driven by human activity and environmental changes. By drawing parallels between the ocean's challenges and familiar human illnesses, we can better understand the urgency of treating these issues and how vital it is to restore the ocean's health. Much like the human body, the ocean is an interconnected system—when one part suffers, the rest begins to fail. This chapter summarizes the different harmful effects due to global change and anthropogenic stressors mentioned along the book.

Temperature Rise and the Parallel with Human Fever

When a person develops a fever, it is a sign that the body is fighting an infection. The increase in body temperature helps speed up metabolic processes and create an environment less suitable for harmful bacteria and viruses. However, if a fever becomes too high or lasts too long, it can cause severe damage to the body's tissues and organs. This analogy of fever is useful when thinking about what is happening to the ocean due to global warming.

The ocean's temperature is rising steadily, like a patient experiencing a prolonged fever. The ocean absorbs about 90% of the excess heat generated by greenhouse gases, which causes the surface waters to warm. Initially, this heat can speed up certain biological processes—plankton grow faster, metabolic rates of marine organisms increase, and some species may temporarily thrive. However, just like in the human body, prolonged heating or heatwaves push the ocean's ecosystems into dangerous territory.

Warmer waters affect coral reefs, disrupt migration patterns of marine species, alter the timing of breeding seasons, and force species to relocate to cooler areas. Just as a fever can cause fatigue, confusion, and dehydration in humans, rising ocean temperatures create stress across marine ecosystems. If the ocean continues to warm, these critical systems will collapse, much like organs failing in the human body during a high fever.

El Niño and La Niña: The Ocean's Seasonal Flu

In humans, the seasonal flu is an illness that many face annually, characterized by cycles of illness and recovery. Though the flu can be disruptive, it is typically temporary, and the body eventually recovers. The ocean has its own version of seasonal flu through the alternating cycles of El Niño and La Niña events. These phenomena are part of the larger El Niño-Southern Oscillation (ENSO) cycle, which causes significant shifts in weather patterns and ocean conditions, particularly in the Pacific Ocean, with ripple effects across the globe.

El Niño is like the flu's feverish stage. During El Niño events, warm water pools in the central and eastern Pacific, disrupting the normal upwelling of nutrient-rich cold water along the coasts of South America. This "warm phase" creates severe disturbances in marine ecosystems. Fish stocks drop as the lack of nutrients in the water causes food chains to collapse, leaving both marine predators and human fisheries without their primary food sources. El Niño also leads to more intense storms, flooding in some areas, and droughts in others, much like the way the flu causes a range of symptoms, from fever to fatigue to body aches.

La Niña, on the other hand, is the opposite phase and represents the cooling down period after a bout of flu. During La Niña events, cooler-than-average sea surface temperatures prevail in the same regions. This tends to restore some balance, but the rapid cooling and increased storm activity also create their own set of challenges, much like how the body recovers from the flu, only to be left weakened and vulnerable for a period of time.

These oscillations are part of a natural cycle, much like the flu season, but climate change is intensifying both El Niño and La Niña events, making them more frequent and severe. This intensification is pushing marine ecosystems to the limit, disrupting seasonal breeding, feeding patterns, and even the global weather systems that humans rely on for agriculture and water supplies. These cycles, like flu outbreaks, can be managed but not prevented, yet human interference is making their effects far worse.

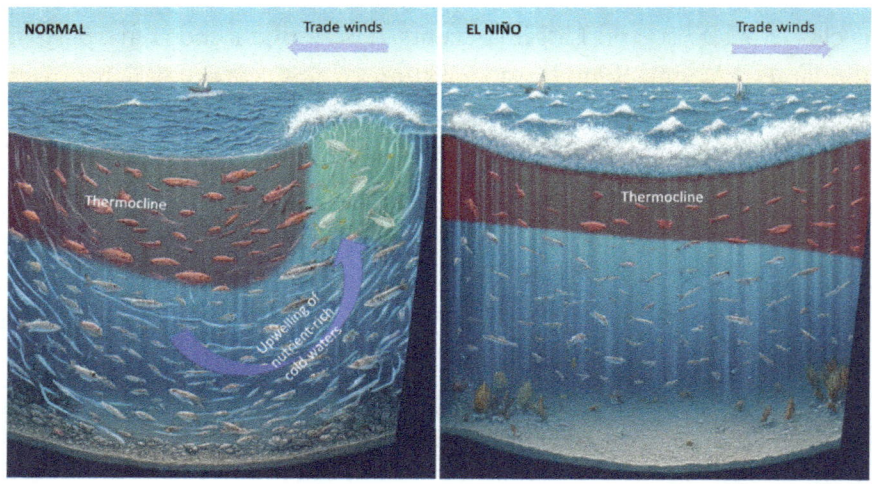

Normal (left) and El Niño conditions (right). Note how the upwelling of nutrient rich waters is prevented during El Niño events.

Skin Diseases of the Ocean: Pollution and the Degradation of Surface Layers

Just as human skin is vulnerable to infections, irritants, and disease, the surface layers of the ocean—where critical exchanges of gases, nutrients, and heat occur—are exposed to a growing list of environmental threats. Human skin acts as a barrier to protect the body from harmful external substances, but when it becomes damaged, infections and diseases can set in. Similarly, the ocean's "skin," or the surface waters, is exposed to pollutants, climate stressors, and physical changes that disrupt its natural balance.

One of the most severe "skin diseases" affecting the ocean is plastic pollution. Every year, millions of tons of plastic waste enter the ocean, where it breaks down into smaller particles called microplastics. These microplastics, much like harmful bacteria on human skin, infiltrate the ocean's surface layers, damaging marine life and ecosystems. Marine organisms, from fish to seabirds, ingest microplastics, mistaking them for food. This disrupts feeding patterns, poisons marine species, and introduces toxic chemicals into the food chain—chemicals that eventually make their way back to humans.

Oil spills are another devastating affliction. When oil coats the surface of the ocean, it blocks sunlight and reduces the ocean's ability to exchange gases with the atmosphere, much like how human skin becomes clogged or damaged by toxins. Without sunlight penetrating the water, photosynthetic organisms like phytoplankton cannot produce the oxygen that supports marine life. In addition, oil spills coat marine animals' bodies, disrupting their ability to regulate temperature, breathe, or find food. These disasters are akin to severe burns or infections in human skin, where healing is slow, painful, and often incomplete.

Another significant threat is the rise of HABs, which are triggered by nutrient pollution, particularly from agricultural runoff. These blooms act like infections in the ocean's skin, growing uncontrollably and releasing toxins that kill fish, marine mammals, and even birds. The blooms also create dead zones—areas where oxygen levels are so low that marine life cannot survive. These blooms are much like human skin diseases that become inflamed and infected, causing widespread damage if not treated.

Cardiovascular Disease of the Ocean: The Slowing of the Gulf Stream

The human heart is essential for circulating blood throughout the body, ensuring that oxygen and nutrients reach every organ and tissue. When the heart's blood flow is blocked or weakened, a heart attack can occur, leading to organ failure and death. In the ocean, as we already mentioned, the equivalent of the heart is the Gulf Stream, a powerful ocean current that circulates warm water from the tropics to the North Atlantic, driving the global climate system and regulating temperature and weather patterns.

Recent research suggests that the Gulf Stream is weakening, primarily due to the melting of polar ice and the influx of fresh water into the North Atlantic. This fresh water reduces the density of the ocean water, disrupting the normal sinking and rising currents that power the Gulf Stream. Just as a blocked artery can prevent blood from reaching the heart, the weakening of this vital current threatens to "stall" the ocean's circulation system.

If the Gulf Stream stops, the consequences will be catastrophic. Northern Europe, which benefits from the warmth brought by the Gulf Stream, would experience a sharp drop in temperature, leading to colder, harsher winters. Meanwhile, tropical regions could become even hotter, destabilizing ecosystems and human populations. This breakdown in circulation is similar to how a heart attack prevents blood from reaching critical areas of the body, leading to the collapse of entire systems. The Gulf Stream's role in maintaining global climate and marine health is irreplaceable, and its failure would represent a cardiovascular crisis for the planet.

Ocean Cancer: Coral Reef Decline and Ocean Acidification

Cancer in the human body is the uncontrolled growth of cells that invade healthy tissue, disrupting the function of vital organs. In the ocean, coral reefs are facing a similarly invasive threat due to ocean acidification and rising temperatures. Ocean acidification, caused by the absorption of excess carbon dioxide, lowers the pH of seawater, making it more difficult for marine organisms like corals, mollusks, and shellfish to form their calcium carbonate skeletons and shells. This is akin to how cancer eats away at healthy tissue in the human body. As the acid levels rise, coral reefs weaken and begin to erode, leading to the collapse of these vital ecosystems.

The decline of coral reefs has a cascading effect throughout the ocean. Coral reefs provide shelter, breeding grounds, and food for a vast array of marine life. As they disappear, the species that depend on them also decline, much like how organs fail when cancer spreads throughout the body. If ocean acidification continues unchecked, entire marine ecosystems could collapse, leading to a loss of biodiversity that would have devastating consequences for both the ocean and humanity.

Immune System Failure: Loss of Biodiversity

The human body's immune system protects against disease, infection, and harmful invaders. When the immune system is weakened—by stress, poor nutrition, or illness—the body becomes more vulnerable to infections and long-term health problems. In the ocean, biodiversity serves as the immune system. A diverse marine ecosystem is resilient, able to withstand environmental changes, recover from disturbances, and defend against invasive species. However, much like a weakened immune system, the ocean's biodiversity is in sharp decline, leaving ecosystems vulnerable to collapse.

Overfishing, habitat destruction, and climate change are rapidly depleting the ocean's biodiversity. Species that play critical roles in maintaining ecosystem balance, such as predators or primary producers, are disappearing. Without these key species, the intricate web of life that supports marine ecosystems begins to unravel. Much like a body weakened by disease, the ocean is becoming less capable of recovering from environmental shocks, such as extreme weather events or pollution.

This loss of biodiversity is akin to immune system failure in humans. Once the ocean loses its ability to bounce back from stress, it becomes increasingly vulnerable to invasive species, harmful algal blooms, and other threats. Restoring biodiversity through conservation efforts, sustainable fishing practices, and habitat protection is essential to ensuring that the ocean's immune system remains strong and capable of defending against future challenges.

The ocean's illnesses, like those of the human body, require diagnosis, treatment, and care. From its fevered waters and skin diseases to its cardiovascular crisis and cancerous coral decline, the ocean is showing clear signs of distress. However, just as human bodies have the ability to heal with proper care, so too can the ocean recover—if we act swiftly and decisively.

CHAPTER 14

OCEANIC HEALING: TREATING THE SEA'S AILMENTS AS DOCTORS OF THE PLANET

When the human body falls ill, we turn to a range of treatments—both modern medicines and holistic therapies—to restore balance, repair damage, and heal. We treat infections with antibiotics, manage chronic conditions with lifestyle changes, and nurture mental and emotional well-being with a combination of practices, from meditation to nutrition. In much the same way, the ocean, too, is suffering from a range of ailments brought on by human activity—pollution, overfishing, climate change, and habitat destruction. Just as we act as healers for our own bodies, it is our responsibility to act as doctors for the ocean, diagnosing its illnesses, understanding their causes, and applying treatments that will allow it to heal and thrive once again.

Becoming Doctors for the Ocean: A Holistic Approach to Ocean Health

When treating chronic conditions like heart disease or diabetes, modern medicine prescribes medications to manage symptoms, but true healing often requires lifestyle changes—better nutrition, regular exercise, stress management, and emotional support. The body is an interconnected system, and healing must address not just the symptoms, but the underlying imbalances that led to illness in the first place. If we are to heal the ocean, we must adopt a similar approach—one that recognizes the ocean as a complex, interconnected system and seeks not just to manage symptoms, but to address the root causes of its decline. This means becoming doctors for the ocean, diagnosing its illnesses, understanding their underlying causes, and applying treatments that are both immediate and long-term, both technological and holistic.

1. Diagnosing the Ocean's Illnesses: Just as doctors rely on tests, scans, and patient histories to diagnose human illnesses, scientists and environmentalists use data, satellite imagery, and field observations to assess the health of the ocean. Rising sea temperatures, shrinking fish populations, increasing acidification, and the spread of plastics and pollutants are clear indicators that the ocean is in distress.

Diagnosing the ocean's diseases requires us to understand how these symptoms are interconnected. For example, coral bleaching is not just about warming waters—it is also linked to overfishing, which reduces the biodiversity that helps coral reefs recover, and pollution, which weakens coral resilience. Dead zones, where oxygen levels drop too low for most marine life, are caused by nutrient runoff from agriculture and urban areas, but they are worsened by overfishing and climate change. The diagnosis is clear: the ocean's illnesses are complex, and no single treatment will cure it

2. Applying Targeted Treatments to Restore Balance: In human medicine, once we diagnose a disease, we apply targeted treatments—whether through surgery, medication, or other interventions. Similarly, targeted actions can help heal specific ocean ecosystems. Marine protected areas, for instance, act like bandages, offering safe zones where fish populations can recover, ecosystems can rebuild, and biodiversity can thrive without the pressures of fishing and industrial exploitation.

Regenerative efforts like coral reef restoration act as the ocean's version of physical therapy, helping weakened ecosystems regain strength. Projects that replant mangroves and seagrass beds serve as lung and liver transplants for the ocean, restoring critical habitats that filter toxins and stabilize coastlines. These targeted interventions are essential for addressing localized damage, but they are not enough on their own—just as medication may relieve symptoms but not cure the underlying condition.

Healing the Earth. A.I.-generated image

3. A Holistic Approach for Treating the Whole Ocean: There is no magical pill. To truly heal the ocean, we must go beyond localized treatments and adopt a holistic approach that addresses the root causes of its decline. This requires rethinking how we interact with the ocean at every level—how we fish, how we manage waste, how we produce energy, and how we structure our economies. Just as holistic health care for humans integrates physical, emotional, and environmental factors, a holistic approach to ocean health requires addressing the interconnected systems that support marine life.

- **Reducing Carbon Emissions**: Climate change is a primary driver of ocean illness. Reducing carbon emissions is akin to cutting out harmful substances from our diet—essential for preventing further harm. A global commitment to transitioning away from fossil fuels and reducing greenhouse gas emissions is the most significant step we can take to restore the ocean's health.

- **Ending Overfishing and Restoring Fisheries**: Overfishing has depleted fish populations and thrown marine food webs out of balance. By reducing fishing pressure, implementing sustainable practices, and allowing fish stocks to recover, we give the ocean a chance to heal itself, just as a body needs time to recover after surgery or illness.

- **Pollution Control**: Just as the human body struggles to detoxify when overwhelmed by pollutants, the ocean is drowning in plastic, chemical waste, and agricultural runoff. Implementing stricter waste management systems, banning single-use plastics, and developing more efficient recycling and disposal methods are necessary steps in removing these toxins from the ocean's system.

- **Ecological Regeneration**: Much like holistic health approaches that promote regenerative healing—through proper diet, stress management, and restorative practices—ecological regeneration can bring the ocean back to life. This includes reforestation of mangroves and seagrass meadows, restoring wetlands, and rebuilding coral reefs, all of which serve as the lungs, kidneys, and skin of the ocean's body.

4. Nurturing the Ocean's Well-Being: In human medicine, healing often involves a balance of active treatments and long-term lifestyle changes. For the ocean, this means finding ways to live in harmony with it. Promoting sustainable fisheries, reducing plastic use, and encouraging renewable energy sources are all essential parts of creating a healthier future for the ocean. Just as we are encouraged to lead healthier lives by managing stress, eating well, and staying active, we must develop a sustainable relationship with the ocean that allows it to flourish in the long term.

We must also nurture the emotional and cultural connections we have with the ocean, recognizing that it is not just a resource, but a vital, living system that sustains all life on Earth. This emotional connection is essential for building the collective determination needed to drive the changes necessary to heal the ocean. Indigenous and holistic worldviews, which see humans as part of a larger ecological system, offer valuable insights into how we can live in balance with the ocean rather than dominating or exploiting it.

Becoming Healers for the Ocean's Future

Just as we take responsibility for our own health, we must now take responsibility for the health of the ocean. We are not just its inhabitants; we are its caretakers. The ocean's illnesses are not inevitable—they are the result of human actions, and they can be healed through human effort. Like skilled doctors, we must combine immediate interventions with long-term lifestyle changes, treating both the symptoms and the root causes of its ailments.

The ocean has an incredible capacity to heal if given the chance. Marine life is resilient, ecosystems can regenerate, and waters can cleanse themselves—if we act quickly and decisively. But this healing requires a collective effort: to become the doctors the ocean desperately needs, diagnosing its diseases, applying treatments, and nurturing it back to health, not just for our own sake, but for the sake of all life on Earth.

As we heal the ocean, we heal ourselves. Our futures are inextricably linked, and just as we value the health of our bodies, we must learn to cherish the health of the ocean, the vast body of water that sustains us all.

CHAPTER 15

THE OCEAN AS A LIVING BEING

The ocean, akin to the human body, is far more than the sum of its parts. It breathes, circulates, protects, senses, and adapts—functioning as a living, interconnected system that sustains life on Earth. Understanding the ocean as a living organism means seeing it not just as a vast body of water but as a dynamic system where every component, from the tiniest microbe to the largest whale, plays a vital role in maintaining its balance. The ocean is Earth's great living system, a body with its own rhythms, signals, and processes that work in harmony to create a thriving, self-sustaining whole.

At the heart of this vast organism is its metabolism—an unceasing flow of energy. Phytoplankton, like the mitochondria of the ocean, convert sunlight into energy that powers marine life. They fuel ecosystems in the same way that energy powers the human body. This energy does not stay isolated—it flows, it moves, it is transformed. Zooplankton graze on phytoplankton, fish eat the zooplankton, and marine predators feed on the fish, cycling energy through the marine food web just as the body's cells, tissues, and organs share and utilize nutrients. Each player in this complex web contributes to the metabolic life of the ocean, ensuring that nutrients are not lost but reused, recycled, and passed on, keeping the ocean's body in motion.

Like the human body's circulatory system, the ocean is crisscrossed by currents that act as arteries and veins, carrying heat, nutrients, and organisms to all corners of the globe. The Gulf Stream, the Antarctic

Circumpolar Current, and deep-water flows are its lifeblood, distributing the warmth that regulates the Earth's climate and delivering vital nutrients to sustain ecosystems. As blood nourishes every cell in the body, these currents nourish every ecosystem, from the vibrant coral reefs to the remote polar seas. Their rhythmic flow ensures that even the most distant and isolated parts of the ocean are connected, just as every organ in the body is part of the whole.

The ocean is also equipped with a sensory system. Its creatures are finely attuned to the world around them, reacting to changes in temperature, salinity, pressure, and chemical signals in ways that mirror the human nervous system's responses to stimuli. Schools of fish, for instance, move as one, a flash of collective decision-making and reflex that mirrors how neurons communicate to trigger muscle movement. Whales use echolocation, sensing their environment with the same precision as our ears interpret sound waves. Coral reefs, sensitive to changes in water quality, temperature, and acidity, act like the skin, reacting to external stressors and signaling deeper imbalances in the ocean's health. Each organism, whether it migrates, adapts, or communicates, plays a part in this vast sensory network, interpreting the ocean's changing conditions and responding in ways that maintain the overall balance of the marine environment.

But like any organism, the ocean is not invulnerable. It has its own immune defenses, capable of repairing damage and fending off threats. Coral reefs secrete mucus to ward off harmful bacteria, while plankton blooms react to nutrient surges, absorbing excess compounds and stabilizing water chemistry. These defensive mechanisms operate quietly in the background, much like the immune system in humans, working to maintain homeostasis, heal wounds, and restore balance. Yet, just as human immune systems can be overwhelmed by chronic stress or illness, the ocean's defenses are faltering under the strain of pollution, overfishing, and climate change. Coral bleaching, dead zones, and declining fish stocks are symptoms of a body under siege, struggling to fend off external threats that it was never evolved to face.

Perhaps most striking is the ocean's capacity for long-term memory and adaptation. Deep in the seabed, carbon and nutrients are sequestered, stored in the marine equivalent of the subconscious, where they slowly shape the future of the planet's climate and life. The deep ocean operates like the depths of the mind, with processes hidden from view but critical for the health of the whole. Carbon sequestration, nutrient cycling, and deep-sea currents act as the ocean's subconscious memory, where lessons from the past—such as historical climate shifts—are stored and slowly processed. The ocean remembers, adapts, and evolves, much like the human brain consolidates experiences and shapes future responses based on past learnings.

A.I. representation of the ocean as a living being

To view the ocean as a living being is to recognize its fragility as well as its resilience. Just as the human body requires care, nourishment, and protection to function properly, the ocean demands the same. It thrives when its ecosystems are in balance, when its currents flow freely, when its species can adapt and thrive in clean, healthy waters. But when this balance is disrupted—through overexploitation, pollution, or climate change—the entire system is put at risk. Much like in the human body, when one system fails, the consequences cascade, affecting the entire organism.

This understanding of the ocean as a living entity is not just a metaphor but a new way of seeing our relationship with the planet. We are part of the Earth's body, not separate from it. The health of the ocean is directly tied to our own well-being. The atmosphere we breathe, the climate we live in, and the food we eat are all tied to the health of the ocean's "body." Just as we would not knowingly harm our own bodies, we must reconsider the ways in which we treat the ocean—seeing it not as an inexhaustible resource to be exploited but as a living, breathing system that requires protection, respect, and care.

The ocean, like the human body, has a remarkable capacity for recovery. Given the chance, it can heal. Coral reefs can regenerate, fish populations can rebound, and nutrient cycles can stabilize. But this recovery depends on us recognizing that the ocean is not merely a resource but a living being, one that we are intrinsically connected to and responsible for. To preserve the health of the ocean is to preserve our own health—because the ocean's body and our own are, in many ways, one and the same.

www.ingramcontent.com/pod-product-compliance
Lightning Source LLC
Chambersburg PA
CBHW070144230526
45471CB00002B/508